# FUNDAMENTALS OF STEELMAKING

# Fundamentals
# of Steelmaking

E.T. TURKDOGAN

THE INSTITUTE OF MATERIALS

Book 614
First published in 1996 by
The Institute of Materials
1 Carlton House Terrace
London SW1Y 5DB

ISBN 0 901716 80 4

Typeset by Dorwyn Ltd,
Rowlands Castle, Hants, UK

Printed and bound in the UK at
The University Press, Cambridge

# Contents

# Acknowledgements

The author is indebted to his wife Myra Andrée Marcelle for editing, typing and proof reading the manuscript with patience and diligence.

I wish to express my appreciation to the research personnel at the U.S. Steel Technical Centre (i) for their collaboration in acquiring for me large volumes of heat logs on steelmaking from USS plants, (ii) for the frequent use of their library facilities and (iii) to the librarian Joy Richardson for being so helpful in my search for information on technical literature.

# Preface

During the past two decades many advances have been made in the implementation of new technologies in the steel industry. These technical advances have been reflected also on the courses given on process metallurgy at universities and technical colleges. This text book therefore is intended for the university graduate students, research scientists and engineers specialising in ferrous pyrometallurgy and for use as a source of reference in the technical training courses given to the steel plant engineers and plant operators.

During the past three score years the manner of applying the principles of thermodynamics and physical chemistry has gone through many phases of development, in the search for a better understanding of the operation and control of the pyrometallurgical processes. Although essentially nothing was known about the thermodynamic properties of slags in the early 1930s, it was nevertheless recognised that, because of the complex ionic nature of molten slags, the classical mass action law could not be used in representing the equilibrium constants of slag-metal reactions. This problem was somewhat circumvented by adopting an empirical approach to the representation of the slag composition as though the slags were made up of a mixture of stoichiometric compounds of various assumed silicates, phosphates and free oxides. This empirical concept is attributed to the early pioneering work of Schenck in his studies of the Bessemer and open-hearth steelmaking reactions.

Numerous attempts had been made in the past, following the work of Herasymenko in the late 1930s, to represent the equilibrium constants of slag-metal reactions in terms of the concentrations of assumed ionic species. Although the derivation of reaction equilibria relations from ionic structural models are of some theoretical interest, their applications to pyrometallurgical reactions in multicomponent systems have not been particularly rewarding.

Studies of thermodynamic properties of metals and slags, initiated primarily by Chipman, Richardson and their co-workers in the late 1940s and 1950s, are still an ongoing subject of research projects. Complementary to these experimental studies, there have been several compilations of thermochemical data made notably by Kelly, Kubaschewski and Hultgren. We are also greatly indebted to Wagner and Darken for their valuable theoretical contributions to the application of the principles of thermodynamics in the study of systems pertinent to various aspects of high temperature technology.

Reflecting upon the many facets of these previous endeavours, I have now reached the conclusion that the equilibrium states of slag-metal reactions, pertinent to the steelmaking conditions, can be quantified in much simpler forms in terms of the mass concentrations of elements dissolved in the steel and of oxides in the slag. In the first two chapters of the book, the principles of thermochemistry and rate phenomena are given in a condensed form (mostly definition of terms and basic equations) that is quite adequate for a comprehensive description of the fundamentals of steelmaking processes and related subjects. Critically reviewed numerical data on the physicochemical properties of gases, iron alloys and slags are compiled in Chapters 3, 4 and 5 which are needed in the formulation of steelmaking reactions. The second half of the book is devoted to (i) the compilation of reassessed equilibrium data on gas–slag–metal reactions, (ii) the discussion of oxygen steelmaking processes and evaluation of the state of reactions in the furnace at tap, (iii) the primary aspects of steel refining in the ladle and (iv) some aspects of reactions occurring during solidification of steel in continuous casting.

May, 1995                                                      E.T. Turkdogan

# CHAPTER 1

# Thermochemistry and Thermodynamics

Selected thermochemical and thermodynamic functions are presented in this chapter in a condensed form without dwelling on their derivations which are well documented in many textbooks. For an in-depth study of thermodynamics, references may be made to the original teachings of Gibbs,[1] the classic texts of Lewis and Randall,[2] later rewritten by Pitzer and Brewer,[3] and of Guggenheim,[4] a textbook on the physical chemistry of metals by Darken and Gurry,[5] an introduction to metallurgical thermodynamics by Gaskell,[6] and many others.

## 1.1 IDEAL GAS LAWS

Definitions of the ideal gas laws are given below as a preamble to statements on the concepts of thermochemistry and thermodynamics.

A relation between temperature, pressure and composition of the system is an equation of state. It was through experimental observations that the equations of state for gases were formulated in the early days of scientific discoveries.

Boyle's law (1662): For a given mass of gas at constant temperature, the gas pressure is inversely proportional to its volume, or

$$\text{pressure} \times \text{volume} = PV = \text{constant} \tag{1.1}$$

Charles's law (1787): At constant volume the pressure exerted on a given mass of gas is a linear function of temperature.

Gay–Lussac's law (1802): This is analogous to Charles' law, but stated in a different form, thus for a given mass of gas at constant pressure, the volume is a linear function of temperature.

$$V_t = V_o(1 + \alpha t) \tag{1.2}$$

where $V_o$ is the volume at $t = 0°C$ and $\alpha$ the coefficient of volume expansion. At relatively high temperatures and low pressures, for ideal gases $\alpha = 1/273.16(°C)$. Therefore for an ideal gas,

$$V_t = V_o \left(1 + \frac{t}{273.16}\right) = V_o \left(\frac{273.16 + t}{273.16}\right) \tag{1.3}$$

The sum $273.16 + t(°C) = T$ is the absolute temperature scale K (Kelvin), which is abbreviated to $T(K) = 273 + t(°C)$ for elevated temperatures.

Avogadro's law (1811): At constant temperature and pressure, equal volumes of all gases contain the same number of molecules. One g-molecule (abbreviated mol) of an ideal gas, containing $6.025 \times 10^{23}$ molecules, occupies a volume of 22.414 litre at 1 atmosphere and 0°C.

### 1.1.1 THE GAS CONSTANT

A gas which obeys the simple gas laws is called an ideal gas satisfying the following relation derived from equations (1.1) and (1.3).

$$PV = nRT \tag{1.4}$$

where n is the number of mols and R the universal molar gas constant.

For one mol of an ideal gas at 273.16 K and *l* atm pressure, the value of the molar gas constant is

$$R = \frac{1 \times 22.414}{273.16} = 0.08205 \; l \; \text{atm mol}^{-1}\text{K}^{-1} \tag{1.5}$$

For pressure in Pa ($= \text{J m}^{-3}$) and volume in $\text{m}^3$,

$$R = \frac{1.01325 \times 10^5 \times 22.414 \times 10^{-3}}{273.16} = 8.314 \; \text{J mol}^{-1}\text{K}^{-1} \tag{1.6}$$

### 1.1.2 GAS PARTIAL PRESSURE, DALTON'S LAW (1801)

In a gas mixture containing $n_1, n_2, n_3 \ldots$ number of mols of gases occupying a volume $V$ at a total pressure $P$, the partial pressures of the constituent gaseous species are as given below.

$$p_1 = \frac{n_1}{n_1 + n_2 + n_3 + \ldots} \times P$$

$$p_2 = \frac{n_2}{n_1 + n_2 + n_3 + \ldots} \times P$$

$$P = p_1 + p_2 + p_3 + \ldots \tag{1.7}$$

Equations for constant pressure, volume and temperature:
The following equations are for a given mass of gas.

| | | |
|---|---|---|
| Constant pressure (isobaric) | $\dfrac{V_1}{T_1} = \dfrac{V_2}{T_2} = \dfrac{V_3}{T_3} \ldots$ | (1.8) |
| Constant volume (isochoric) | $\dfrac{P_1}{T_1} = \dfrac{P_2}{T_2} = \dfrac{P_3}{T_3} \ldots$ | (1.9) |
| Constant temperature (isothermal) | $P_1V_1 = P_2V_2 = P_3V_3 \ldots$ | (1.10) |

## 1.1.3 NONIDEAL GASES

The behaviour of gases becomes nonideal at low temperatures and high pressures. Since the metallurgical processes are at elevated temperatures and ordinary pressures, the equation of state for nonideal gases will not be discussed here.

## 1.2 THE FIRST LAW OF THERMODYNAMICS

The first law of thermodynamics is based on the concept of conservation of energy. When there is interaction between systems, the gain of energy of one of the systems is equal to the loss of the other system. For example, the quantity of heat required to decompose a compound into its elements is equal to the heat generated when that compound is formed from its elements.

## 1.2.1 ENERGY

In 1851 Lord Kelvin defined the term 'energy' as follows:

> The energy of a material system is the sum, expressed in mechanical units of work, of all the effects which are produced outside the system when the system is made to pass in any manner from the state in which it happens to be to a certain arbitrarily fixed initial state.

That is, there is no such thing as absolute energy but only relative energy; only change in energy with change of state can be measured, with reference to a standard state.

## 1.2.2 ENTHALPY (HEAT CONTENT)

The internal energy of a system includes all forms of energy other than the kinetic energy. Any exchange of energy between a system and its surroundings, resulting from a change of state, is manifested as heat and work.

When a system expands against a constant external pressure $P$, resulting in an increase of volume $\Delta V$, the work done by the system is

$$w = P\Delta V = P(V_B - V_A) \tag{1.11}$$

Since this work is done by the system against the surroundings, the system absorbs a quantity of heat q and the energy E of the system increases in passing from state A to state B.

$$\Delta E = E_B - E_A = q - P\Delta V = q - P(V_B - V_A) \qquad (1.12)$$

Upon re-arranging this equation, we have

$$(E_B + PV_B) - (E_A + PV_A) = q \qquad (1.13)$$

The quantity $E + PV$ is represented by a single symbol $H$, thus

$$\Delta H = q = (E_B + PV_B) - (E_A + PV_A) \qquad (1.14)$$

The function $H$ is known as enthalpy or heat content.

## 1.2.3 HEAT CAPACITY

The heat capacity of a substance is defined as the quantity of heat required to raise the temperature by one degree. The heat capacity of 1 g of a substance is called the specific heat. The heat capacity of 1 g-molecule (abbreviated as mol) is called the molar heat capacity.

For an ideal gas the difference between the molar heat capacities at constant pressure, $C_P$, and constant volume, $C_V$, is equal to the molar gas constant.

$$C_P - C_V = R \qquad (1.15)$$

Because of experimental convenience, the heat capacity is determined under conditions of constant pressure (usually atmospheric).

The change in enthalpy of a system with temperature at constant pressure is by definition the heat capacity of the system.

$$C_P = \left(\frac{\partial H}{\partial T}\right)_P \qquad (1.16)$$

Integration gives the change in enthalpy with change in temperature at constant pressure.

$$\Delta H = H_{T_2} - H_{T_1} = \int_{T_1}^{T_2} C_p dT \qquad (1.17)$$

Above 298 K, the temperature dependence of $C_p$ may be represented by

$$C_p = a + bT - cT^{-2} \qquad (1.18)$$

where the coefficients, $a$, $b$ and $c$ are derived from $C_p$ calorimetric measurements at different temperatures.

$$\Delta H = \int_{298}^{T} (a + bT - cT^{-2})\, dT \qquad (1.19)$$

In recent compilations of thermochemical data, the $\Delta H$ values are tabulated at 100 K intervals for the convenience of users.

### 1.2.4 STANDARD STATE

The enthalpy is an extensive property of the system, and only the change in heat content with change of state can be measured. A standard reference state is chosen for each element so that any change in the heat content of the element is referred to its standard state, and this change is denoted by $\Delta H°$.

The natural state of elements at 25°C and 1 atm pressure is by convention taken to be the reference state. On this definition, the elements in their standard states have zero heat contents.

The heat of formation of a compound is the heat absorbed or evolved in the formation of 1 g-mol of the compound from its constituent elements in their standard states, denoted by $\Delta H°_{298}$.

### 1.2.5 ENTHALPY OF REACTION

The change of enthalpy accompanying a reaction is given by the difference between the enthalpies of the products and those of the reactants.

For an isobaric and isothermal reaction,

$$A + B = C + D \tag{1.20}$$

the enthalpy change is given by

$$\Delta H° = (\Delta H°_C + \Delta H°_D) - (\Delta H°_A + \Delta H°_B) \tag{1.21}$$

By convention, $\Delta H$ is positive (+) for endothermic reactions, i.e. heat absorption, and $\Delta H$ is negative (–) for exothermic reactions, i.e. heat evolution.

Temperature effect:

$$\Delta H°_T = \Sigma \Delta H°_{298} \text{ (products)} - \Sigma \Delta H°_{298} \text{ (reactants)}$$

$$+ \int_{298}^{T} [\Sigma C_p \text{ (products)} - \Sigma C_p \text{ (reactants)}]dT \tag{1.22}$$

$$\Delta H°_T = \Delta H°_{298} + \int_{298}^{T} (\Delta C_p)dT \tag{1.23}$$

1.2.6 HESS'S LAW

In 1840 Hess defined the law of constant heat summation as 'the heat change in a chemical reaction is the same whether it takes place in one or several stages.' This law of Hess is in fact a direct consequence of the law of conservation of energy.

1.2.7 SPECIAL TERMS OF HEAT OF REACTION

| | | | |
|---|---|---|---|
| Enthalpy or heat of formation | $Fe + \frac{1}{2}O_2$ | $\rightarrow$ | $FeO$ |
| Heat of combustion | $C + O_2$ | $\rightarrow$ | $CO_2$ |
| Heat of decomposition | $2CO$ | $\rightarrow$ | $C + CO_2$ |
| Heat of calcination | $CaCO_3$ | $\rightarrow$ | $CaO + CO_2$ |
| Heat of fusion (melting) | Solid | $\rightarrow$ | Liquid |
| Heat of sublimation | Solid | $\rightarrow$ | Vapour |
| Heat of vaporisation | Liquid | $\rightarrow$ | Vapour |
| Heat of solution | $Si(l)$ | $\rightarrow$ | [Si] (diss.in Fe) |

1.2.8 ADIABATIC REACTIONS

When a reaction occurs in a thermally insulated system, i.e. no heat exchange between the system and its surroundings, the temperature of the system will change in accordance with the heat of reaction.

As an example, let us consider the internal oxidation of unpassivated direct reduced iron (DRI) in a stockpile, initially at 25°C. The enthalpy of reaction at 298 K is

$$Fe + \frac{1}{2}O_2 \rightarrow FeO, \Delta H^\circ_{298} = -267 \text{ kJ mol}^{-1} \tag{1.24}$$

The heat balance calculation is made for 1000 kg Fe in the stockpile with 150 kg FeO formed in oxidation. The heat absorbed by the stockpile is $(150 \times 10^3/72) \times 267$ kJ and the temperature rise is calculated as follows:

$Q \qquad = [n_{Fe} (C_P)_{Fe} + n_{FeO} (C_P)_{FeO}](T - 298)$

$n_{Fe} \qquad = 17,905$ g-mol for 1000 kg Fe

$n_{FeO} \qquad = 2087.7$ g-mol for 150 kg FeO

$C_P (Fe) \quad = 0.042$ kJ mol$^{-1}$K$^{-1}$

$C_P (FeO) = 0.059$ kJ mol$^{-1}$K$^{-1}$

$\qquad\qquad \therefore Q = 557,416 = (752 + 123) (T - 298)$

With this adiabatic reaction, the stockpile temperature increases to $T = 935$ K (662°C).

The moisture in the stockpile will react with iron and generate $H_2$ which will ignite at the elevated stockpile temperature. This has been known to happen when DRI briquettes were had not adequately passivated against oxidation.

## 1.3 THE SECOND LAW OF THERMODYNAMICS

The law of dissipation of energy states that all natural processes occurring without external interference are spontaneous (irreversible processes). For example, heat conduction from a hot to a cold part of the system. The spontaneous processes cannot be reversed without some change in the system brought about by external interference.

### 1.3.1 ENTROPY

The degree of degradation of energy accompanying spontaneous, hence irreversible, processes depends on the magnitude of heat generation at temperature $T$ and temperatures between which there is heat flow.
The quantity $q/T$ is a measure of degree of irreversibility of the process, the higher the quantity $q/T$, the greater the irreversibility of the process. The quantity $q/T$ is called the increase in entropy. In a complete cycle of all reversible processes the sum of the quantities $\Sigma q/T$ is zero.

The thermodynamic quantity, entropy $S$, is defined such that for any reversible process taking place isothermally at constant pressure, the change in entropy is given by

$$dS = \frac{dH}{T} = \frac{C_P}{T} \ dT = C_P d \ (\ln T) \tag{1.25}$$

## 1.4 THE THIRD LAW OF THERMODYNAMICS

The heat theorem put forward by Nernst (1906) constitutes the third law of thermodynamics: 'the entropy of any homogeneous and ordered crystal-line substance, which is in complete internal equilibrium, is zero at the absolute zero temperature.' Therefore, the integral of equation (1.25) given above has a finite value at temperature T as shown below.

$$S_T = \int_0^T C_p d \ (\ln T) \tag{1.26}$$

The entropy of reaction is

$$\Delta S = \Sigma S(\text{products}) - \Sigma S(\text{reactants}) \tag{1.27}$$

and entropy of fusion at the melting point $T_m$

$$\Delta S_m = \frac{\Delta H_m}{T_m} \tag{1.28}$$

## 1.5 GIBBS FREE ENERGY

From a combined form of the first and second laws of thermodynamics, Gibbs derived the free energy equation for a reversible process at constant pressure and temperature.

$$G = H - TS \tag{1.29}$$

The Gibbs free energy is also known as the chemical potential.

When a system changes isobarically and isothermally from state $A$ to state $B$, the change in the free energy is

$$G_B - G_A = \Delta G = \Delta H - T\Delta S \tag{1.30}$$

During any process which proceeds spontaneously at constant pressure and temperature, the free energy of the system decreases. That is, the reaction is thermodynamically possible when $\Delta G < O$. However, the reaction may not proceed at a perceptible rate at lower temperatures, if the activation energy required to overcome the resistance to reaction is too high. If $\Delta G > O$, the reaction will not take place spontaneously.

As in the case of enthalpy, the free energy is a relative thermodynamic property with respect to the standard state, denoted by $\Delta G°$.

The variation of the standard free energy change with temperature is given by

$$\Delta G_T^\circ = \Delta H_{298}^\circ + \int_{298}^{T} \Delta C_p dT - T\Delta S_{298}^\circ - T \int_{298}^{T} \Delta C_p d\,(\ln T) \tag{1.31}$$

### 1.5.1 GENERALISATION OF ENTROPY OF REACTION

1. When there is volume expansion accompanying a reaction, i.e. gas evolution, at constant pressure and temperature the entropy change is positive, hence $\Delta G$ decreases with an increasing temperature.

$$C + CO_2 = 2CO$$

$$\Delta G° = 166,560 - 171.0T \text{ J}$$

2. When there is volume contraction, i.e. gas consumed in the reaction, at constant pressure and temperature the entropy change is negative, hence $\Delta G$ increases with an increasing temperature.

$$H_2 + \tfrac{1}{2}S_2 = H_2S$$

$$\Delta G° = -91,600 + 50.6T \text{ J}$$

3. When there is little or no volume change the entropy change is close to zero, hence temperature has little effect on $\Delta G$.

$$C + O_2 = CO_2$$

$$\Delta G^\circ = -395,300 - 0.5T \text{ J}$$

### 1.5.2 THE STANDARD FREE ENERGIES OF FORMATION OF COMPOUNDS

For many reactions, the temperature dependence of $\Delta H^\circ$ and $\Delta S^\circ$ are similar and tend to cancel each other, thus the nonlinearity of the variation of $\Delta G^\circ$ with temperature is minimised. Using the average values of $\Delta \tilde{H}^\circ$ and $\Delta \tilde{S}^\circ$, the free energy equation is simplified to

$$\Delta G^\circ = \Delta \tilde{H}^\circ - \Delta \tilde{S}^\circ T \qquad (1.32)$$

The standard free energies of reactions encountered in ferrous metallurgical processes can be computed using the free energy data listed in Table 1.1.

## 1.6 EMPIRICAL CORRELATIONS

### 1.6.1 HEAT CAPACITIES OF SOLID ELEMENTS AND SIMPLE COMPOUNDS

Over the years attempts have been made in pursuit of rationales for composition dependence of the thermochemical properties of substances. A classical example is that of Dulong and Petit (1819) who noted that for most solid elements, the atomic heat capacity at room temperature and atmospheric pressure is essentially constant within the range $25.5 \pm 1.7$ J atom$^{-1}$K$^{-1}$, which is very close to $3R = 24.94$ J atom$^{-1}$K$^{-1}$. There are a few exceptions: the elements beryllium, boron, carbon and silicon have lower atomic heat capacities, and elements cerium, gadolinium, potassium and rubidium higher atomic heat capacities.

Dulong & Petit's rule was subsequently extended to simple compounds by various investigators, ultimately leading to what is known as Kopp's rule (1865): 'the molar heat capacity of a solid compound is approximately equal to the sum of the atomic heat capacities of its constituent elements.'

The combination of the two rules leads to the following approximation in terms of the molar gas constant $R$.

$$\frac{\text{molar } C_P}{\text{number of atoms}} = \tilde{C}_P \approx 3R \qquad (1.33)$$

It should be emphasised that this approximation applies only to simple compounds, e.g. oxides, sulphides, nitrides, carbides, halides and intermetallic compounds containing less than 4 or 5 atoms per mol formula.

**Table 1.1**   The standard free energies of formation of selected compounds from compiled thermochemical data.[7,8]

Notations: (g) gas, (l) liquid, (s) solid, (d) decomposition, (m) melting, (v) vaporisation

| | $\Delta G^\circ = \Delta \bar{H}^\circ - \Delta \tilde{S}^\circ T$ | | | |
|---|---|---|---|---|
| | $-\Delta \bar{H}^\circ$ kJ mol$^{-1}$ | $-\Delta \tilde{S}^\circ$ J mol$^{-1}$K$^{-1}$ | ± kJ | Temp.Range °C |
| $2Al(l)+3/2\,O_2(g) = Al_2O_3(s)$ | 1683.2 | 325.6 | 8 | 659–1700 |
| $Al(l)+1/2\,N_2(g) = AlN(s)$ | 328.3 | 115.5 | 4 | 659–1700 |
| $4Al(l)+3C(s) = Al_4C_3(s)$ | 266.5 | 96.2 | 8 | 659–1700 |
| $2B(s)+3/2\,O_2(g) = B_2O_3(l)$ | 1228.8 | 210.0 | 4 | 450–1700 |
| $B(s)+1/2\,N_2(g) = BN(s)$ | 253.6 | 89.5 | 2 | 25–900 |
| $4B(s)+C(s) = B_4C(s)$ | 79.5 | 9.2 | 4 | 25–900 |
| $C(s)+2H_2(g) = CH_4(g)$ | 91.0 | 110.7 | 2 | 25–2000 |
| $C(s)+1/2\,O_2(g) = CO(g)$ | 114.4 | −85.8 | 2 | 25–2000 |
| $C(s)+O_2(g) = CO_2(g)$ | 395.3 | −0.5 | 2 | 25–2000 |
| $Ca(l)+1/2\,O_2(g) = CaO(s)$ | 900.3 | 275.1 | 6 | 850–1487v |
| $Ca(l)+1/2\,S_2(g) = CaS(s)$ | 548.1 | 103.8 | 4 | 850–1487v |
| $3CaO(s)+Al_2O_3(s) = Ca_3Al_2O_6(s)$ | 16.3 | −26.4 | 8 | 25–1535d |
| $12CaO(s)+7Al_2O_3(s) = Ca_{12}Al_{14}O_{33}(s)$ | 73.1 | −207.5 | 8 | 25–1455m |
| $CaO(s)+Al_2O_3(s) = CaAl_2O_4(s)$ | 19.1 | −17.2 | 8 | 25–1605m |
| $CaO(s)+CO_2(g) = CaCO_3\ (s)$ | 161.3 | 137.2 | 4 | 25–880d |
| $2CaO(s)+SiO_2(s) = Ca_2SiO_4(s)$ | 118.8 | −11.3 | 10 | 25–1700 |
| $CaO(s)+SiO_2(s) = CaSiO_3(s)$ | 92.5 | 2.5 | 12 | 25–1540m |
| $0.947Fe(s)+1/2\,O_2(g) = Fe_{0.947}O\ (s)$ | 263.7 | 64.3 | 4 | 25–1371m |
| $Fe(l)+1/2\,O_2(g) = FeO\ (l)$ | 225.5 | 41.3 | 4 | 1537–1700 |
| $3Fe(s)+2\,O_2(g) = Fe_3O_4(s)$ | 1102.2 | 307.4 | 4 | 25–1597m |
| $2Fe(s)+3/2\,O_2(g) = Fe_2O_3(s)$ | 814.1 | 250.7 | 4 | 25–1500 |
| $Fe(s)+1/2\,S_2(g) = FeS(s)$ | 154.9 | 56.9 | 4 | 25–988m |
| $H_2(g)+1/2\,O_2(g) = H_2O(g)$ | 247.3 | 55.9 | 1 | 25–2000 |
| $H_2(g)+1/2\,S_2(g) = H_2S(g)$ | 91.6 | 50.6 | 1 | 25–2000 |
| $3/2\,H_2(g)+1/2\,N_2(g) = NH_3(g)$ | 53.7 | 32.8 | 0.5 | 25–2000 |
| $Mg(g)+1/2\,O_2(g) = MgO(s)$ | 759.4 | 202.6 | 10 | 1090–2000 |
| $Mg(g)+1/2\,S_2(g) = MgS(s)$ | 539.7 | 193.0 | 8 | 1090–1700 |
| $2MgO(s)+SiO_2(s) = Mg_2SiO_4(s)$ | 67.2 | 4.3 | 8 | 25–1898m |
| $MgO(s)+SiO_2(s) = MgSiO_3(s)$ | 41.1 | 6.1 | 8 | 25–1577m |
| $MgO(s)+CO_2(g) = MgCO_3(s)$ | 116.3 | 173.4 | 8 | 25–402d |
| $Mn(s)+1/2\,O_2(g) = MnO(s)$ | 384.8 | 73.1 | 4 | 25–1244 |
| $Mn(l)+1/2\,O_2(g) = MnO(s)$ | 397.0 | 81.1 | 4 | 1244–1700 |
| $Mn(s)+1.2\,S_2(g) = MnS(s)$ | 277.9 | 64.0 | 4 | 25–1244 |
| $Mn(l)+1/2\,S_2(g) = MnS(s)$ | 290.1 | 71.9 | 4 | 1244–1530m |
| $Mn(l)+1/2\,S_2(g) = MnS(l)$ | 262.6 | 64.4 | 4 | 1530–1700 |
| $MnO(s)+SiO_2(s) = MnSiO_3(s)$ | 28.0 | 2.8 | 12 | 25–1291m |
| $1/2S_2(g)+O_2(g) = SO_2(g)$ | 361.7 | 72.7 | 0.5 | 25–1700 |
| $Si(l)+1/2\,O_2(g) = SiO(g)$ | 154.7 | −52.5 | 12 | 1410–1700 |
| $Si(s)+O_2(g) = SiO_2(s)$ | 907.1 | 175.7 | 12 | 400–1410 |
| $Si(l)+O_2(g) = SiO_2(s)$ | 952.7 | 203.8 | 12 | 1410–1723m |

## 1.6.2 HEAT CAPACITIES OF POLYMERIC SUBSTANCES

From a detailed study of the heat capacity data for complex polymeric compounds, the Author[9] noted that there was indeed a rationale for

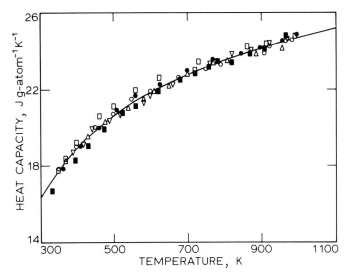

Fig. 1.1   Heat capacities of crystalline minerals and aluminosilicate glasses in terms of J g-atom$^{-1}$K$^{-1}$, using the experimental data of Krupka *et al.*[10] From Ref. 9.

composition dependence of the heat capacity. Based on the heat capacity data of Krupka *et al.*[10] for several aluminosilicates,* it was found that the heat capacity per atom is a single function of temperature as shown in Fig. 1.1. At temperatures of 300 to 400 K, $\bar{C}_P$ values are much below 3R, indicating that Kopp's rule does not hold for complex compounds. The mythical 3R value happens to be at temperatures of 950 to 1050 K for which no particular significance can be attached. The curve in Fig. 1.1 may be represented by the following equation.

$\bar{C}_P$ (J atom$^{-1}$K$^{-1}$) =

$$49.75 - 0.0127T + 180 \times 10^3 T^{-2} - 555 T^{-1/2} + 5 \times 10^{-6} T^2 \qquad (1.34)$$

The molar heat capacities of polymeric substances at 1000 K, including molybdates, titanates, tungstates and so on, are seen from the plot in Fig. 1.2 to be proportional to the number of atoms per mol formula of the compounds, with a slope of $3R$ as anticipated from the data in Fig. 1.1.

| | | | |
|---|---|---|---|
| *Albite: | NaAlSi$_3$O$_8$ | Muscovite: | KAl$_2$(AlSi$_3$O$_{10}$)(OH)$_2$ |
| Orthoclase: | KAlSi$_3$O$_8$ | Pyrophillite: | Al$_2$Si$_4$O$_{10}$(OH)$_2$ |
| Anorthite: | CaAl$_2$Si$_2$O$_8$ | Grossular: | Ca$_3$Al$_2$Si$_3$O$_{12}$ |

### 1.6.3 HEAT CONTENTS AT MELTING POINTS

When there is no allotropic phase transformation, the integration of equation (1.34) gives the following expression for the heat content relative to 298 K.

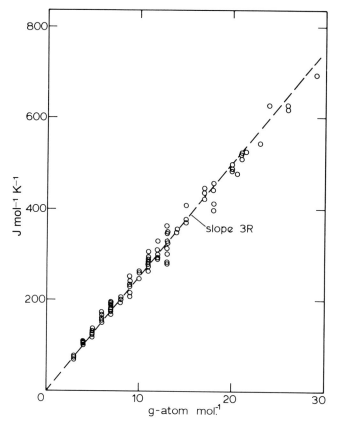

Fig. 1.2   Molar heat capacities of polymeric substances (network formers) at 1000 K related to g-atom per mol; the broken line is drawn with a slope of 3$R$. From Ref. 9.

$$H_T^\circ - H_{298}^\circ \ (\text{J atom}^{-1}) =$$
$$5460 + 49.75T - 6.35 \times 10^{-3}T^2 - 180 \times 10^3 T^{-1} - 1110T^{1/2}$$
$$+ 1.67 \times 10^{-6}T^3 \qquad (1.35)$$

Using the available thermochemical data compiled by Barin et al.[11,12] the values of $H_T^\circ - H_{298}^\circ$ per g-atom, minus the enthalpies of phase transformations, are computed for solid polymeric substances at their melting points; these are plotted in Fig. 1.3. The full-line curve calculated from equation (1.35) is in general accord with the enthalpy data for a variety of crystalline substances which have network-type structures, with the number of atoms per mol formula being in the range 3–26 and melting points 600 to 2200 K. Within the scatter of the data, the relation in Fig. 1.3 may be simplified to a linear equation for temperatures above 1100 K, with $\Delta H_t$ per g-atom added for the phase transformations in the solid state.

$$H_T^\circ - H_{298}^\circ \ (\text{kJ atom}^{-1}) = 0.03T - 15.6 + \Delta H_t \qquad (1.36)$$

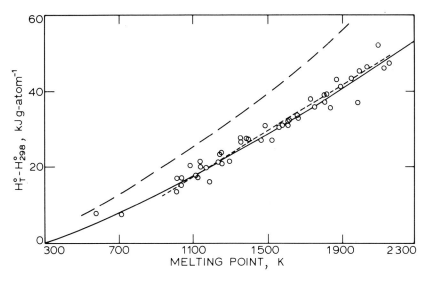

Fig. 1.3 Heat contents of polymeric substances (network formers) at their melting points; the upper curve — — — is approximate for the liquid phase. From Ref. 9.

## 1.6.4 HEAT OF MELTING

Since thermodynamic quantities are functions of similar state properties, one thermodynamic quantity may be related to another. With this thought in mind, in Fig. 1.4, $\Delta H_m$ is plotted against $H_T^\circ - H_{298}^\circ$ at the melting points of polymeric substances. With the exception of the datum point for silica ($\Delta H_m = 3.19$ kJ atom$^{-1}$), there is a fair correlation which may be represented by the following approximation.

$$\Delta H_m \text{ (kJ atom}^{-1}) \approx 0.35(H_{Tm}^\circ - H_{298}^\circ) - 2.5 \qquad (1.37)$$

## 1.7 THERMODYNAMIC ACTIVITY

The combined statement of the first and second laws for a system doing work only against pressure gives the following thermodynamic relation.

$$dG = VdP - SdT \qquad (1.38)$$

At constant temperature $\Delta G = VdP$ and for 1 mol of an ideal gas $V = RT/P$; with these substituted in equation (1.38) we obtain

$$dG = RT \ \frac{dP}{P} = RT \ d \ln P \qquad (1.39)$$

Similarly, for a gas mixture

$$dG_i = RT \ d \ln p_i \qquad (1.40)$$

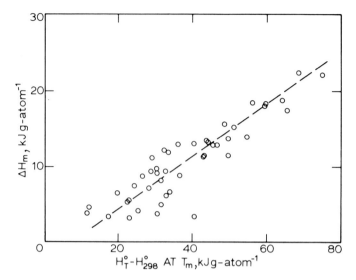

Fig. 1.4   Enthalpy of fusion related to heat contents of polymeric substances at their melting points. From Ref. 9.

where $p_i$ is the partial pressure of the $i$th species in the gas mixture, and $\bar{G}_i$ partial molar free energy.

In a homogeneous liquid or solid solution, the thermodynamic activity of the dissolved element is defined by the ratio

$$a_i = \left( \frac{\text{vapour pressure of component } (i) \text{ in solution}}{\text{vapour pressure of pure component}} \right)_T \qquad (1.41)$$

In terms of solute activity, the partial molar free energy equation is

$$d\bar{G}_i = RT \, d \ln a_i \qquad (1.42)$$

Integration at constant temperature gives the relative partial molar free energy in solution

$$\bar{G}_i = RT \ln a_i \qquad (1.43)$$

### 1.7.1 TEMPERATURE EFFECT ON ACTIVITY

In terms of the relative partial molar enthalpy and entropy of solution

$$\bar{G}_i = \bar{H}_i - \bar{S}_i T \qquad (1.44)$$

which gives

$$\ln a_i = \frac{\bar{H}_i}{RT} - \frac{\bar{S}_i}{R} \qquad (1.45)$$

or

$$\log a_i = \frac{\bar{H}_i}{2.303 \, RT} - \frac{\bar{S}_i}{2.303 \, R} \qquad (1.46)$$

1.7.2 SOLUTIONS

A solution is a homogeneous gas, liquid or solid mixture, any portion of which has the same state properties. The composition of gas solution is usually given in terms of partial pressures of species in equilibrium with one another under given conditions. For liquid solutions, as liquid metal and slag, the composition is given in terms of the molar concentrations of components of the solution.

The atom or mol fraction of the component *i* in solution is given by the ratio

$$N_i = \frac{n_i}{\Sigma n}$$

where $n_i$ is the number of g-atoms or mols of component *i* per unit mass of solution, and $\Sigma n$ the total number of g-atoms or mols. Since the metal and slag compositions are reported in mass percent, $n_i$ per 100 g of the substance is given by the ratio

$$n_i = \frac{\%i}{M_i}$$

where $M_i$ is the atomic or molecular mass of the component *i*.

Noting that the atomic mass of iron is 55.85g, the atom fraction of solute *i* in low alloy steels is given by a simplified equation

$$N_i = \frac{\%i}{M_i} \times 0.5585 \qquad (1.47)$$

In low alloy steelmaking, the composition of slag varies within a relatively narrow range, and the total number of g-mol of oxides per 100 g of slag is within the range $\Sigma n = 1.6 \pm 0.1$. With this simplification, the mol fraction of the oxide in the slag is given by

$$N_i = \frac{\%i}{1.6 \, M_i} \qquad (1.48)$$

1.7.3 MOLAR QUANTITIES OF SOLUTIONS

The molar free energy of mixing (solution), $G^M$, enthalpy of mixing $H^M$, and entropy of mixing, $S^M$, are given by the following summations:

$$G^M = N_1 \bar{G}_1 + N_2 \bar{G}_2 + N_3 \bar{G}_3 + \ldots \qquad (1.49a)$$

$$H^M = N_1 \bar{H}_1 + N_2 \bar{H}_2 + N_3 \bar{H}_3 + \ldots \qquad (1.49b)$$

$$S^M = N_1 \bar{S}_1 + N_2 \bar{S}_2 + N_3 \bar{S}_3 + \ldots \qquad (1.49c)$$

For binary systems, a graphical method is used to evaluate the partial molar quantities from the molar quantity of solution by drawing tangents to the curve for the molar quantity *G* as demonstrated in Fig. 1.5.

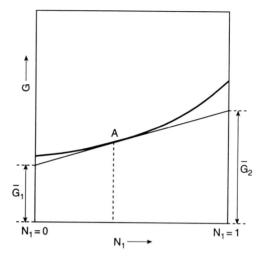

Fig. 1.5   Graphical method of evaluating the partial molar quantities from the molar quantity of a binary system.

### 1.7.4 IDEAL SOLUTIONS – RAOULT'S LAW (1887)

The solutions are said to be ideal, if the activity is equal to the mol or atom fraction of the component $i$ in solution,

$$a_i = N_i \qquad (1.50)$$

A thermodynamic consequence of Raoult's law is that the enthalpy of mixing for an ideal solution, $H^{M,id}$, is zero. Substituting $a_i = N_i$ and $H^{M,id} = 0$ in the free energy equation gives for the entropy of formation of an ideal solution.

$$S^{M,id} = -R(N_1 \ln N_1 + N_2 \ln N_2 + N_3 \ln N_3 + \ldots) \qquad (1.51)$$

### 1.7.5 NONIDEAL SOLUTIONS

Almost all metallic solutions and slags exhibit nonideal behaviour. Depending on the chemical nature of the elements constituting a solution, the activity vs composition relation deviates from Raoult's law to varying degrees, as demonstrated in Fig. 1.6 for liquid Fe–Si and Fe–Cu systems at 1600°C.

### 1.7.6 ACTIVITY COEFFICIENT

The activity coefficient of solute $i$ is defined by the ratio

$$\gamma_i = \frac{a_i}{N_i} \qquad (1.52)$$

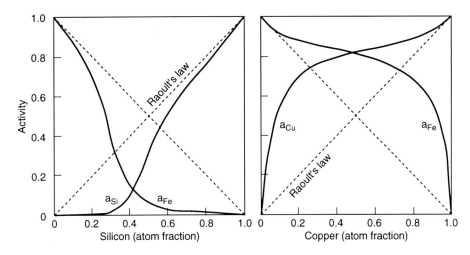

Fig. 1.6   Activities in liquid Fe–Si and Fe–Cu alloys at 1600°C showing strong negative and positive departures from Raoult's law.

If the activity is relative to the pure component $i$, it follows from Raoult's law that as $N_i \rightarrow 1$, $\gamma_i \rightarrow 1$.

*1.7.6a Henry's law for dilute solutions*

In infinitely dilute solutions, the activity is proportional to the concentration

$$a_i = \gamma_i^\circ N_i \tag{1.53}$$

The approach to Henry's law at dilute solutions is demonstrated in Fig. 1.7 for the activity of carbon (relative to graphite) in austenite at 1000°C. The austenite containing 1.65%C ($N_C = 0.072$) is saturated with graphite at 1000°C for which the carbon activity is one relative to graphite.

Since Henry's law is valid at infinite dilution only, the ratio $\gamma_i/\gamma_i^\circ$ is used as a measure of departure from Henry's law for finite solute contents in dilute solutions. For solute concentration in terms of mass percent, Henry's activity coefficient is defined by the ratio

$$f_i = \frac{\gamma_i}{\gamma_i^\circ} \tag{1.54}$$

such that $f_i \rightarrow 1$ when $\%i \rightarrow 0$.

*1.7.6b Interaction coefficients*

Over several mass percentages of the solute, the composition dependence of activity coefficient $f_i$ of solute $i$ in binary systems is represented by the following relation in terms of mass %i.

$$\log f_i = e^i{}_i \, [\%i] \tag{1.55}$$

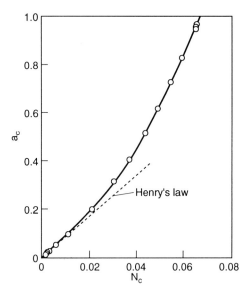

Fig. 1.7   Activity of carbon in austenite (relative to graphite) at 1000°C, demonstrating deviation from Henry's law.

where e is called the solute interaction coefficient. For multi-component solutions, the following summation is used

$$\log f_i = e^i{}_i\,[\%i] + \Sigma e^j{}_i\,[\%j] \tag{1.56}$$

where $e^j{}_i$ is the effect of the alloying element $j$ on the activity coefficient of solute $i$.

### 1.7.7 CONVERSION FROM ONE STANDARD STATE TO ANOTHER

In steelmaking processes we are concerned with reactions involving dissolved elements at low concentrations in liquid steel. Therefore, it is convenient to express the solute activity relative to Henry's law. The free energy change accompanying the isothermal transfer of solute from state A to state B is

$$\Delta G = RT \ln \frac{a_B}{a_A}$$

Taking a pure component for state $A$, i.e. $a_A = 1$ and Henry's law for state $B$, i.e. $a_B = \gamma^\circ_i N_i$

$$\Delta G = RT \ln(\gamma^\circ_i N_i)$$

For one mass percent solute in iron,

$$N_i \approx \frac{0.5585}{M_i}$$

where $M_i$ is the atomic mass (g) of the solute. Assuming that Henry's law holds at 1 mass %, the standard free energy of solution of pure component $i$ in iron at 1 mass % is

$$\Delta G_s = RT \ln \left( \frac{0.5585}{M_i} \gamma_i^\circ \right) \tag{1.57}$$

Since Henry's law is valid at infinite dilution only, appropriate correction must be made for nonideal behaviour when using the above equation.

## 1.8 REACTION EQUILIBRIUM CONSTANT

A reaction at constant temperature and pressure,

$$mM + nN = uU + vV$$

is accompanied by a change in the free energy of the system thus,

$$\Delta G = (u\bar{G}_U + v\bar{G}_V) - (m\bar{G}_M + n\bar{G}_N)$$

As shown earlier, the relative partial molar free energy of solution is

$$\bar{G}_i = G_i^\circ + RT \ln a_i$$

Inserting this equality in the above equation gives

$$\Delta G = [u(G_U^\circ + RT \ln a_U) + v(G_V^\circ + RT \ln a_V)]$$
$$- [m(G_M^\circ + RT \ln a_M) + n(G_N^\circ + RT\, a_N)]$$

At equilibrium $\Delta G = 0$, therefore, in terms of the activities of reactants and products, the standard free energy change accompanying the reaction is

$$\Delta G^\circ = (uG_U^\circ + vG_V^\circ - mG_M^\circ - nG_N^\circ) = - RT \ln \frac{(a_U)^u (a_V)^v}{(a_M)^m (a_N)^n} \tag{1.58}$$

The equilibrium constant is

$$K = \frac{(a_U)^u (a_V)^v}{(a_M)^m (a_N)^n} \tag{1.59}$$

which inserted in equation (1.58) gives

$$\Delta G^\circ = - RT \ln K \tag{1.60}$$

### 1.8.1 EFFECT OF TEMPERATURE

Since $\Delta G^\circ$ is a function of temperature only, the equilibrium constant is also a function of temperature only.

$$\ln K = -\Delta G^\circ / RT$$

In terms of enthalpy and entropy changes

$$\ln K = -\frac{\Delta H}{RT} + \frac{\Delta S}{R} \tag{1.61}$$

In exothermic reactions $\Delta H$ is negative, therefore the equilibrium constant $K$ decreases with an increasing temperature. The temperature effect on $K$ is the opposite in endothermic reactions.

### 1.9 PHASE RULE

For the state of equilibrium in the system influenced by temperature, pressure and composition only, from thermodynamic considerations Gibbs derived the following relation, known as the phase rule

$$f = n - p + 2 \tag{1.62}$$

where $f$ is the degrees of freedom (i.e. number of variables which define the system), $n$ the number of components and $p$ the number of phases.

For a single component system such as $H_2O$, $n = 1$. For a single phase water, $p = 1$ hence $f = 2$; that is, both temperature and pressure can be changed arbitrarily. For water-water vapour, water-ice or ice-water vapour equilibrium $p = 2$ and $f = 1$; in this univariant equilibrium either temperature or pressure is the independent variable. When the three phases (ice, water and water vapour) are in equilibrium, $f = 0$; this is an invariant system at a particular temperature and pressure which defines the triple point.

The pressure vs temperature relation is shown in Fig. 1.8 for a single component system. If the triple point is above atmospheric pressure, the solid sublimes without melting. For example, solid $CO_2$ sublimes at $-78.5°C$ at 1 atm; melting occurs at $-56.4°C$ at 5.11 atm.

### 1.10 PHASE EQUILIBRIUM DIAGRAMS

For ordinary pressures of metallurgical interest, mostly at about 1 atm, the solid-liquid phase equilibrium is not influenced perceptibly by pressure; the only remaining independent variables of the system are temperature and composition, consequently for essentially constant pressure the phase rule is simplified to

$$f = n - p + 1 \tag{1.63}$$

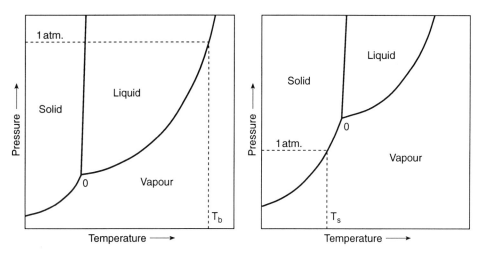

Fig. 1.8   Phase relations for single component systems.

### 1.10.1 BINARY SYSTEMS

When the components *A* and *B* of a binary system are mutually soluble in the liquid and solid state, the phase equilibrium diagram has the simplest form as shown in Fig. 1.9. In this system, the solidification of a melt of composition X occurs within the temperature range $T_1$ and $T_2$ and the compositions of liquid and solid solutions change along the $L_1L_2$ liquidus and $S_1S_2$ solidus curves respectively. At an intermediate temperature $T'$, the ratio $L'X'/X'S'$ gives the amount of solid solution $S'$ relative to the residual liquid of composition $L'$.

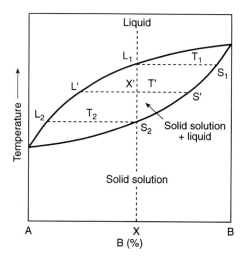

Fig. 1.9   Simple binary phase diagram with complete liquid and solid solutions.

When there is a greater positive departure from ideal behaviour in the solid solution than for the liquid, the phase diagrams are of the forms shown in Figs. 1.10 and 1.11 for the eutectic and peritectic systems with partial solid solutions.

In the binary eutectic system within the composition range $A'$ and $B'$, both $\alpha$ and $\beta$ phases are formed at the last stage of solidification. With three phases present the degree of freedom is zero, hence the eutectic composition and temperature are constant and the solidus line between $A'$ and $B'$ is drawn parallel to the composition abscissa.

In the system shown in Fig. 1.11, there is a peritectic reaction between the solid solution $\alpha$ of composition $C$ and liquid $P$ forming the solid

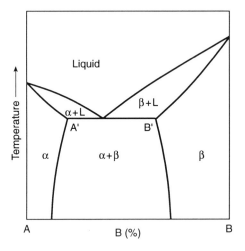

Fig. 1.10   Binary eutectic systems with partial solid solubilities.

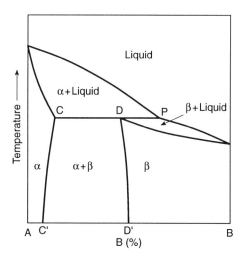

Fig. 1.11   Peritectic reaction in a binary system with partial solubilities.

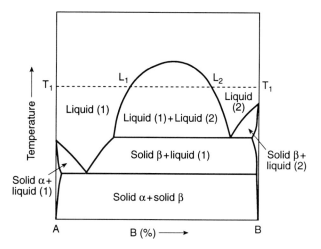

Fig. 1.12   Liquid miscibility gap in a binary system.

solution β of composition D. Within the field CDC'D' there are two solid phases α and β, the compositions of which change with temperature along the solubility curves CC' and DD' respectively.

When there is a strong positive departure from ideal behaviour in the liquid solution, a miscibility gap occurs over some composition and temperature range as drawn in Fig. 1.12. The horizontal line depicting the three-phase equilibrium between solid β, liquid (1) and liquid (2) is called the monotectic invariant. With increasing temperature the mutual solubilities of liquid (1), rich in A, and liquid (2), rich in B, increase and above a particular critical temperature the two liquids become completely miscible. The activity vs composition relation for a system with a miscibility gap is shown in Fig. 1.13. Since the liquids $L_1$ and $L_2$ are in equilibrium, the activity of A will be the same in both phases as indicated by the horizontal line $L_1L_2$.

1.10.2 TERNARY SYSTEMS

In a ternary system there are two degrees of freedom for the two-phase region: two solids or one solid and liquid. The alternative arbitrary changes that can be made are (i) the concentrations of two of the components or (ii) temperature and concentration of one of the components. That is, in a ternary system the two-phase region is depicted by a curved surface in the three dimensional temperature-composition diagram as illustrated in Fig. 1.14. This simple ternary eutectic system is made up from three binary eutectic systems A–B, A–C and B–C. The liquidus surfaces are DFGE for component A, DHJE for component B and JKGE for component C. These three liquidus surfaces merge at the eutectic invariant point E; the

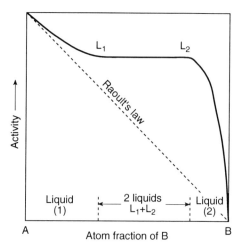

Fig. 1.13   Activity of component A at temperature $T_1$ of the system given in Fig. 1.12; pure liquid A is the standard state.

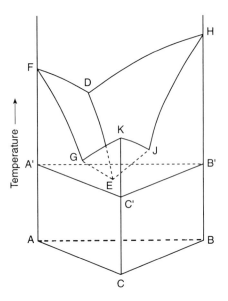

Fig. 1.14   Three dimensional sketch of a ternary eutectic system.

solidus invariant is the surface $A'B'C'$ that is parallel to the base composition triangle $ABC$.

The liquidus surfaces of ternary phase diagrams are depicted in a two-dimensional drawing by projecting the liquidus isotherms onto the composition triangle as shown in Fig. 1.15 for the system $Fe_3O_4$–$SiO_2$–$Al_2O_3$ at 1 atm air.[13] The phase equilibrium relations in ternary systems are described more easily by drawing isothermal sections as in Fig. 1.16 for 1500°C. The dotted lines in two-phase regions delineate liquid

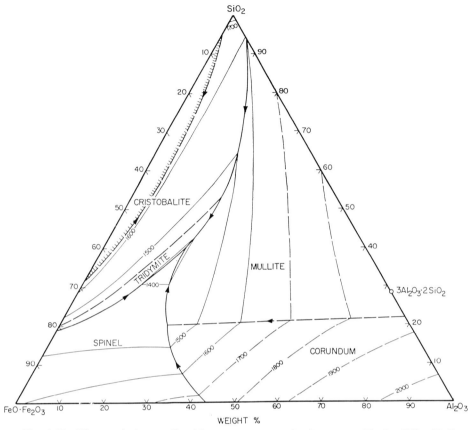

Fig. 1.15   Phase relations at liquidus temperatures in the system Fe$_3$O$_4$–SiO$_2$–Al$_2$O$_3$ in air, based mainly on the data of Muan.[13]

compositions in equilibrium with the solid phase. In the triangular regions there are three phases: two solid phases in equilibrium with liquid of fixed composition for a given isotherm.

## 1.11 Surface Tension (Surface Energy)

The plane of separation of two phases is known as a surface or interface. For a given system at constant temperature and pressure, the increase in free energy per unit increase in the surface area, is the surface tension (surface energy) or interfacial tension $\sigma$

$$\sigma = \left( \frac{\partial G}{\partial A} \right)_{T,P,n_i} \qquad (1.64)$$

where $G$ is energy, $A$ the surface area and $n_i$ the number of mols of the $i$th component.

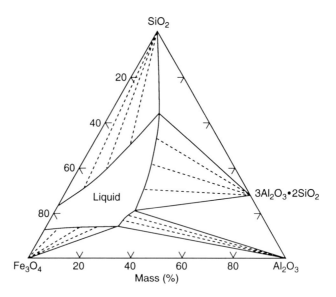

Fig. 1.16  Phase equilibrium relations at 1500°C isotherm.

For a binary system at constant temperature and pressure, the following thermodynamic relation was derived by Gibbs.

$$d\sigma = - kT\Gamma_i \, d(\ln a_i) \tag{1.65}$$

where $a_i$ is the solute activity, $\Gamma_i$ the excess surface concentration of the chemisorbed solute and $k$ the Boltzmann constant $= 1.38 \times 10^{-23}$ J K⁻¹. Rearranging the above equation gives

$$\Gamma_i = - \frac{1}{kT} \frac{d\sigma}{d(\ln a_i)} \tag{1.66}$$

This thermodynamic relation is known as the Gibbs adsorption equation.

Units of $\sigma$:
as surface tension                  Nm⁻¹   Numerical values
as surface energy                   Jm⁻²   are identical

Units of $\Gamma_i$:                             Number of atoms per m².

In the integrated form, the Gibbs adsorption equation becomes

$$\sigma = \sigma_o - kT\Gamma_i \ln a_i \tag{1.67}$$

where $\sigma_o$ is the surface energy of the pure substance in an inert atmosphere.

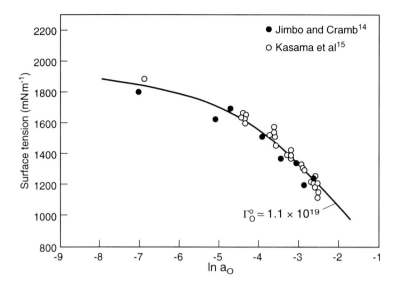

Fig. 1.17   Effect of oxygen on surface tension of liquid iron at 1550°C.

### 1.11.1 CHEMISORBED LAYER

The surface concentration of solute in the chemisorbed layer is determined by drawing tangents to the curve in the plot of $\sigma$ against $\ln a_i$, as shown in Fig. 1.17 for the liquid Fe–O melts at 1550°C. At high solute activities, the curve approaches linearity asymptotically; this limiting slope gives the surface concentration at saturation designated by $\Gamma_i^\circ$. The fraction of sites covered by adsorbed species at constant temperature and solute activity is given by the ratio

$$\theta_i = \left( \frac{\Gamma_i}{\Gamma_i^\circ} \right)_{T,a_i} \tag{1.68}$$

### 1.11.2 LANGMUIR ADSORPTION ISOTHERM

Based on the kinetic theory of gases and the reaction kinetics of adsorption and desorption, Langmuir[16] derived the following relation for the mono-layer chemisorption in a binary system.

$$a_i = \varphi_i \ \frac{\theta_i}{1-\theta_i} \tag{1.69}$$

If there is no interaction between the chemisorbed species, the monolayer is said to be ideal for which the coefficient $\varphi_i$ is independent of coverage of the surface sites.

The heat of adsorption $\Delta H_a$ for a given coverage $\varphi_i$ is obtained from the temperature dependence of $\varphi_i$.

$$\Delta H_a = R \; \frac{d \ln \varphi_i}{d(1/T)} \tag{1.70}$$

According to the surface tension data for the liquid Fe–O system at 1550°C, the monolayer saturation is approached at about $a_O = 0.1$ ($\approx 0.1$ wt.% O) for which $\Gamma_O^\circ \sim 1.1 \times 10^{19}$ O atoms/m². The oxygen adsorption isotherm for liquid iron at 1550°C derived from the surface tension data is plotted in Fig. 1.18. According to this estimate of the adsorption isotherm, the coefficient $\varphi_O$ decreases from 0.014% as $\theta_O \to 0$ to 0.0013% as $\theta_O \to 1$, which indicates strong negative deviation from ideal behaviour in the monolayer.

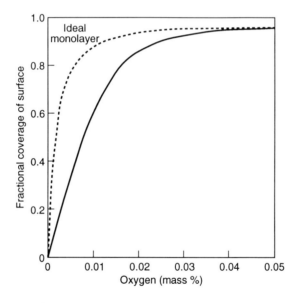

Fig. 1.18   Fraction of surface coverage by chemisorbed oxygen on liquid iron at 1550°C, derived from the surface tension data in Fig. 1.17.

Bernard and Lupis[17] suggested an adsorption model which takes into account interaction forces between adsorbed atoms. On the other hand, Belton[18] assumed the ideal monolayer coverage in deriving adsorption isotherms for various metal-oxygen and sulphur systems from the surface tension data. It is interesting to note that the adsorption isotherms derived on the assumption of an ideal monolayer are similar to those based on the model proposed by Bernard and Lupis. It appears that, within the accuracy limits of the experimental data, surface tensions are not too sensitive to the choice of the adsorption function.

For the limiting case of $\theta_i$ approaching 1, the fraction of sites not occupied, $1-\theta_i$, is inversely proportional to the solute activity

$$1-\theta_i = \frac{\varphi_i}{a_i} \qquad (1.71)$$

With liquid iron this limiting case is approached at $\sim 0.1\%$ O or $\sim 0.01\%$ S.

### 1.11.3 EFFECT OF SURFACE CURVATURE ON SURFACE TENSION

From thermodynamic considerations, Kelvin derived the following relation for a curved surface, here given in terms of the vapour pressure of species $i$,

$$\ln \frac{P_i}{P^*_i} = \left( \frac{2\sigma}{r} \right)\left( \frac{\bar{V}_i}{RT} \right) \qquad (1.72)$$

where $P_i$ is the vapour pressure over a curved surface and $P^*_i$ for a plane surface, $r$ the radius of curvature and $\bar{V}_i$ the relative partial molar volume of the $i$th component in solution.

For bubbles in a liquid, the excess pressure in the bubble is given by

$$\Delta P = P - P^* = \frac{2\sigma}{r} \qquad (1.73)$$

## 1.12 INTERFACIAL TENSION

Interfacial tension between two solids, two dis-similar liquids, or solid and liquid, plays a role to a varying degree of importance in the processing of materials, e.g. ore beneficiation by flotation, sintering, inclusion coalescence in liquid metals, emergence of inclusions from liquid steel to over laying slag and so on.

### 1.12.1 CONTACT ANGLE

Examples of contact angles are illustrated in Fig. 1.19 for solid-liquid and liquid-liquid systems.

The contact angle $\theta$ is defined as the angle included within the liquid phase. When $\theta > 90°$, the system is non-wetting. Wetting of the solid by the liquid occurs when $\theta < 90°$, complete wetting being achieved when $\theta = 0°$. Under equilibrium conditions at the triple point $A$, the surface tension $\sigma_{sg}$ is balanced by the horizontal resolution of tensions $\sigma_{sl}$ and $\sigma_{lg}$ in the opposite direction, thus

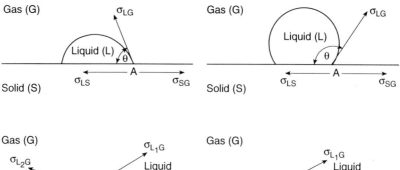

Fig. 1.19   Contact angles at solid–liquid and liquid–liquid interfaces.

solid–liquid:                     $\sigma_{sg} = \sigma_{sl} + \sigma_{lg} \cos \theta$                     (1.74a)

liquid–liquid:         $\sigma_{l_2g} \cos \Phi = \sigma_{l_1g} \cos \alpha + \sigma_{l_1l_2} \cos \beta$         (1.74b)

1.12.2 WORK OF COHESION, WORK OF ADHESION AND SPREADING
COEFFICIENT

The work required to split a column of liquid or solid into two surfaces of
unit area is called the *work of cohesion, $W_c$*.

$$W_c = 2\sigma_{l_1g}$$                     (1.75)

Consider a column of liquid $l_1$, interfacing with another liquid $l_2$ or solid
$s$. Upon separation at this interface, two new surfaces of unit area are
created ($l_1g$ and $l_2g$ or $sg$) and one interface of unit area is lost ($l_1l_2$ or $l_1s$).
The accompanying change of interfacial energy is called the *work of adhesion, $W_a$*.

$$W_a = \sigma_{l_1g} + \sigma_{l_2g} - \sigma_{l_1l_2}$$                     (1.76)

The spreading coefficient $S_p$ of liquid ($l$) on liquid (2) or on a solid is
given by the difference between the work of adhesion and work of
cohesion

$$S_p = W_a - W_c$$

$$= \sigma_{l_2g} - \sigma_{l_1g} - \sigma_{l_1l_2}$$                     (1.77a)

$$= \sigma_{sg} - \sigma_{l_1g} - \sigma_{l_1s}$$                     (1.77b)

The spreading or wetting occurs when $S_p$ is positive. That is, spreading of liquid (1) on particles of liquid (2) or on a solid particle, will occur when

$$\sigma_{l_1 l_2} < \sigma_{l_2 g} - \sigma_{l_1 g} \tag{1.78a}$$

or

$$\sigma_{l_1 s} < \sigma_{sg} - \sigma_{l_1 g} \tag{1.78b}$$

## 1.13 COMPUTER-BASED PACKAGES

The computer-based packages are routinely used in the chemical, metallurgical and many other industries to calculate the reaction equilibria in multiphase systems. This subject is outside the scope of this book; however, a few references are given below on computer-based programmes in relation to metallurgical processes.

a. MTDATA:
The National Physical Laboratory (UK) computer package MTDATA is a thermodynamic database which can be used to determine the equilibrium compositions of multi-component mixtures of gases, liquids and solids. Application of MTDATA to modelling of slag, matte, metal and gas phase equilibria is illustrated in a publication by Dinsdale *et al.*[19]

b. F*A*C*T:
The Facility for the Analysis of Chemical Thermodynamics (F*A*C*T), developed by Thompson *et al.*[20], is an online database computing system. The system performs a wide range of computations from simple gas reactions to heterogeneous equilibria involving up to 12 components and 500 species. Graphical output is provided with programmes which generate Pourbaix diagrams, isothermal predominance diagrams and binary and ternary potential-composition phase diagrams.

c. SOLGASMIX:
This computer package is based on the concept of the Gibbs free energy minimisation which determines reaction equilibria in multiphase metallurgical systems.[21,22].

d. ChemSage:
The computer programme called ChemSage, based on SOLGASMIX Gibbs energy minimiser, is designed to perform three types of thermochemical calculations in complex systems which involves phases exhibiting nonideal mixing properties: (i) calculations of thermodynamic functions, (ii) heterogeneous phase equilibria and (iii) steady-state conditions for the simulation of simple multistage reactors. The module for ther-

modynamic functions calculates specific heat, enthalpy, entropy and free energy with respect to a chosen reference state for a given phase; when the phase is a mixture, the partial quantities of its components are evaluated. Chemical equilibrium calculations can be made for a system which has been uniquely defined with respect to temperature, pressure (or volume) and composition. One of these quantities may be replaced by an extensive property or phase target, as for example the calculation of adiabatic and liquidus temperatures, respectively. The use of this Chem-Sage programme is demonstrated in a paper by Eriksson and Hack.[23]

## REFERENCES

1. *The Collected Works of J. Willard Gibbs*, vols I and II. Longmans, Green, New York, 1928.
2. G.N. LEWIS and M. RANDALL, *Thermodynamics and the free energy of chemical substances*, McGraw-Hill, New York, 1923.
3. K.S. PITZER and L. BREWER, *Thermodynamics – Lewis & Randall*, McGraw-Hill, New York, 1961.
4. E.A. GUGGENHEIM, *Thermodynamics*, Wiley Interscience, New York, 1949.
5. L.S. DARKEN and R.W. GURRY, *Physical chemistry of metals*, McGraw-Hill, New York, 1953.
6. D. GASKELL, *Introduction to metallurgical thermodynamics*, McGraw-Hill, New York, 1973.
7. E.T. TURKDOGAN, *Physical chemistry of high temperature technology*, Academic Press, New York, 1980.
8. E.T. TURKDOGAN, *Ironmaking and Steelmaking*, 1993, **20**, 469.
9. E.T. TURKDOGAN, *Physicochemical properties of molten slags and glasses*, The Metals Society, London, 1983.
10. K.M. KRUPKA, R.A. ROBIE and B.S. HEMINGWAY, *Am. Minerolog.*, 1979, **64**, 86.
11. I. BARIN and O. KNACKE, *Thermochemical properties of inorganic substances*, Springer-Verlag, Berlin, 1973.
12. I. BARIN, O, KNACKE and O. KUBASCHEWSKI, *Thermochemical properties of inorganic substances – Supplement*, Springer-Verlag, Berlin, 1977.
13. A. MUAN, *J. Am. Ceram. Soc.*, 1957, **40**, 121.
14. I. JIMBO and A.W. CRAMB, *Iron Steel Inst. Japan, International*, 1992, **32**, 26.
15. A. KASAMA, A. MCLEAN, W.A. MILLER, Z. MORITA and M.J. WARD, *Can. Met. Quart.*, 1983, **22**, 9.
16. I. LANGMUIR, *J. Am. Chem. Soc.*, 1916, **38**, 2221; 1918, **40**, 1361.
17. G. BERNARD and C.H.P. LUPIS, *Surf. Sci.*, 1974, **42**, 61.
18. G.R. BELTON, *Metall. Trans. B*, 1976, **B7**, 35.
19. A.T. DINSDALE, S.M. HODSON and J.R. TAYLOR, in '3rd International Conference on molten slags and fluxes', p. 246. The Institute of Metals, London, 1989.

20  W.T. THOMPSON, G. ERIKSSON, C.W. BALE and A.D. PELTON, in 'Computerized metallurgical databases symposium', p. 53, eds. J.R. Cuthill, N.A. Gokcen and J.E. Morral. *Metall. Soc. of AIME*, Warrendale, Pa, USA, 1988.

21  G. ERIKSSON and E. ROSEN, *Chem. Scr.*, 1973, **4**, 193.

22  G. ERIKSSON, *Chem. Scr.*, 1975, **8**, 100.

23  G. ERIKSSON and  K. HACK, *Metall. Trans. B.*, 1990, **21B**, 1013.

CHAPTER 2

# Rate Phenomena

There are many different facets of rate phenomena involving homogeneous or heterogeneous chemical reactions, mass transfer via atomic or molecular diffusional processes, viscous flow, thermal and electrical conduction and so on.

In high temperature processes as in pyrometallurgy, the rates of interfacial chemical reactions are in general much faster than the rates of transfer of the reactants and reaction products to and from the reaction site. The formulations of rate equations for transport controlled reactions vary considerably with the physical properties and type of fluid flow over the surface of the reacting condensed phase. Then, there are formulations of rate equations for different regimes of gas bubbles in liquid metal and slag. For a comprehensive discussion of the transport controlled rate phenomena in metallurgical processes, reference may be made to textbooks by Szekely and Themelis,[1] and by Geiger.[2] In section 2.8 references are given to a few publications on the use of computer software in the mathematical and physical modelling of various aspects of the rate phenomena which are encountered in metallurgical processes.

The rate phenomena discussed in this chapter are on chosen subjects which are pertinent to various aspects of ferrous–pyrometallurgical processes, in addition to oxygen steelmaking, ladle refining and degassing.

Selected examples are given of laboratory experiments on the study of reaction rates in simple systems to demonstrate how the rate controlling reaction mechanisms are identified and the rate constants determined.


**2.1 KINETICS OF INTERFACIAL REACTIONS**

In steelmaking and related processes we are concerned with heterogenous reactions involving an interface between two reacting phases, e.g. solid–liquid, solid–gas, liquid–gas and two immiscible liquids (slag–liquid steel).

For the case of a fast rate of transport of reactants and products to and from the reaction site, the rate is controlled by a chemical reaction occurring in the adsorbed layer at the interface. The reaction between adsorbed species $L$ and $M$ on the surface producing product $Q$ occurs via the formation of an activated complex $(LM)$.*

$$L + M \quad = \quad (LM)^* \quad \rightarrow \quad Q \qquad (2.1)$$
$$\left\{ \begin{array}{l} \text{stage I} \\ \text{reactants} \end{array} \right\} \quad \left\{ \begin{array}{l} \text{activated} \\ \text{complex} \end{array} \right\} \quad \left\{ \begin{array}{l} \text{stage II} \\ \text{products} \end{array} \right\}$$

34

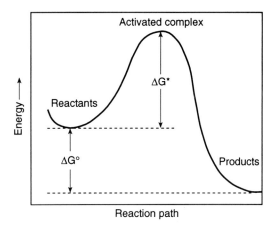

Fig. 2.1  Change in energy profile during the formation and decomposition of the activated complex involved in the reaction.

The theory of the absolute reaction rates is based on the concept of the formation of an activated complex as an intermediate transition state, which has an infinitesimally short life time of the order of $10^{-15}$ second. For an in-depth study of the theory of reaction kinetics, to which outstanding contributions were made by Eyring and co-workers, reference may be made to the classical text books by Glasstone *et al.*[3] and by Hinshelwood.[4]

As illustrated in Fig. 2.1, there is a change in energy profile accompanying the reaction that involves the formation and decomposition of an activated complex. While the change in free energy accompanying reaction (2.1) is $\Delta G° < 0$, the activation energy $\Delta G^* > 0$.

The theory of the absolute rates states that the activated complex is in equilibrium with the reactants for which the equilibrium constant $K^*$ for constant temperature is

$$K^* = \frac{a^*}{a_L a_M} \tag{2.2}$$

where $a$s are the thermodynamic activities.

Next to be considered is the specific rate of decomposition of the activated complex to the overall reaction product $Q$, represented by

$$\frac{dn}{dt} = \left(\frac{kT}{h}\right) \Gamma_o \theta^* \tag{2.3}$$

where $\dfrac{dn}{dt}$ = the reaction rate, mol cm$^{-2}$s$^{-1}$

   $k$ = the Boltzmann constant, $1.380 \times 10^{-23}$ J K$^{-1}$
   $h$ = the Planck constant, $6.626 \times 10^{-34}$ J s
   $T$ = temperature, K
   $\Gamma_o$ = total number of adsorption sites on the surface, $\approx 10^{15}$ mole cm$^{-2}$
   $\theta^*$ = fractional coverage by the activated complex.

For single site occupancy by the activated complex in the adsorbed layer, the activity of the complex is represented by

$$a^* = \varphi^* \frac{\theta^*}{1 - \theta} \tag{2.4}$$

where $\theta$ is the total fractional occupancy of the sites by the adsorbed species and $\varphi^*$ is the activity coefficient of the complex in the chemisorbed layer. Combining equations (2.2), (2.3) and (2.4) gives the rate of forward reaction in terms of the absolute reaction rate theory.

$$R_f = \frac{dn}{dt} = \left( \frac{kT}{h} \right) \Gamma_o \left( \frac{K^*}{\varphi^*} \right) (1 - \theta) \{a_L a_M\} \tag{2.5}$$

The thermodynamics of the chemisorbed layer at the interface, i.e. values of $\Gamma_o$, $K^*$ and $\varphi^*$, are not known, therefore the isothermal rate equation is given in a simplified general form thus

$$R_f = \Phi_f (1 - \theta) \{a_L a_M\} \tag{2.6}$$

where $\Phi_f$ is the isothermal rate constant of the forward reaction.

As the reaction progresses, concentrations of the reactants $L$ and $M$ decrease while the concentration of the product $Q$ increases. Because of these composition changes and the influence of the reverse reaction $Q \rightarrow L + M$, the rate decreases with an increasing reaction time. The rate of the reverse reaction is represented by

$$R_r = - \Phi_r (1 - \theta) \{a_Q\} \tag{2.7}$$

where $\Phi_r$ is the rate constant of the reverse reaction. Therefore, the net overall rate of reaction is

$$\frac{dn}{dt} = \Phi_f (1 - \theta) \{a_L a_M\} - \Phi_r (1 - \theta) \{a_Q\} \tag{2.8}$$

When the rates of forward and reverse reactions are the same, i.e. $dn/dt = 0$, the reaction is said to be at an equilibrium state. It follows that the ratio of the rate constants $\Phi_f/\Phi_r$ is the equilibrium constant of the reaction.

$$K = \frac{\Phi_f}{\Phi_r} = \left( \frac{a_Q}{a_L a_M} \right)_{eq} \tag{2.9}$$

In terms of a single rate constant, the net reaction rate is formulated as

$$\frac{dn}{dt} = \Phi_f (1 - \theta) \{a_L a_M - (a_L a_M)_{eq}\} \tag{2.10}$$

where $(a_L a_M)_{eq}$ is the equilibrium value for the activity $a_Q$ in stage II at any given reaction time.

For a given surface coverage $\theta$, the temperature dependence of the rate constant is represented by

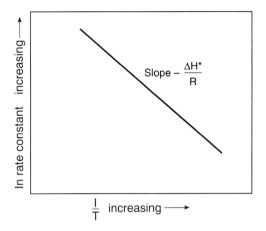

Fig. 2.2  Effect of temperature on rate constant for reaction involving an activated complex.

$$\Phi = [\Phi_o \exp(-\Delta H^*/RT)]_\theta \tag{2.11}$$

where $\Phi_o$ is a pre-exponential constant and $\Delta H^*$ is the heat of activation for the reaction at a given site fillage $\theta$.

Since the activation energy $\Delta G^*$ is always positive, the enthalpy of activation for the reaction is also positive. As shown in Fig. 2.2, in the plot ln $\Phi$ vs $1/T$ the slope of the line gives $\Delta H^*$. It should be noted that since the thermodynamic quantities for the activated complex $(K^*/\varphi^*)$ in the chemisorbed layer are not known, the $\Delta H^*$ derived from the rate measurements over a sufficiently large temperature range, is the apparent heat of activation.

Another aspect of the kinetics of interfacial reactions is the rate of chemisorption or desorption at the reaction surface as given by the following theoretical equation for an uncontaminated surface.[3]

$$\Phi_f = (2\pi M_i RT)^{-1/2} \exp(-\Delta H^*/RT) \tag{2.12}$$

where $M_i$ is the molecular mass of the adsorbed species. This equation is transformed to the following form for the maximum rate of vaporisation, i.e. free vaporisation, from an uncontaminated surface at low pressures

$$Rate_{max} = \frac{p_i}{\sqrt{2\pi M_i RT}} \tag{2.13}$$

where $p_i$ is the vapour pressure in atm. for which the equation is reduced to the following form.

$$Rate_{max} \text{ (g-mol cm}^{-2}\text{s}^{-1}) = 44.3 p_i (M_i T)^{-1/2} \tag{2.14}$$

## 2.1.1 EXAMPLES OF EXPERIMENTS ON RATES OF INTERFACIAL REACTIONS

### 2.1.1a Nitrogen transfer across iron surface

The rate of reaction of gaseous nitrogen with liquid and solid iron, in the presence or absence of surface active alloying elements, has been studied by many investigators since the late 1950s. It was in the late 1960s that the rate controlling reaction mechanism was resolved,[5–7] which was further confirmed in the subsequent experimental work done by Fruehan and Martonik[8] and by Byrne and Belton.[9]

When the rate of reaction is not hindered by slow nitrogen transport to and from the gas-metal interface, the rate of nitrogenation

$$N_2(g) = N^*_2 \rightarrow 2[N] \tag{2.15}$$

is controlled either by the rate of chemisorption or dissociation of $N_2$ molecules on the metal surface. The rate of reverse reaction, i.e. denitrogenation, is of course a second order type with respect to nitrogen dissolved in the metal. The equation below represents the rate of nitrogen transfer from gas to liquid iron

$$\frac{d\,[\%N]}{dt} = \frac{100A}{\rho V}\,\Phi_f\,(1-\theta)\,\left\{p_{N_2} - (p_{N_2})_{eq}\right\} \tag{2.16}$$

where $\rho$ is the density of liquid iron, $A$ the surface area of the melt on which the nitrogen stream is impinging and $V$ the volume of the melt. The rate constant $\Phi_f$, in units of g N cm$^{-2}$min$^{-1}$atm$^{-1}$N$_2$, is for the forward reaction (2.15). The equilibrium partial pressure $(p_{N_2})_{eq}$ corresponding to the nitrogen content of the melt at the reaction time $t$, is that given by the equilibrium constant $K$ for nitrogen solubility.

$$(p_{N_2})_{eq} = [\%N]^2/K \tag{2.17}$$

With this substitution the isothermal rate equation is

$$\frac{d\,[\%N]}{dt} = \frac{100A}{\rho V}\,\Phi_f\,(1-\theta)\,\left\{p_{N_2} - [\%N]^2/K\right\} \tag{2.18}$$

For constant $N_2$ pressure and temperature, the integration of equation (2.18) gives for $\%N = 0$ at $t = 0$,

$$\ln\frac{Kp_{N_2} + [\%N]}{Kp_{N_2} - [\%N]} = 2p_{N_2}\,\frac{100A}{\rho V}\,\Phi_f\,(1-\theta)t \tag{2.19}$$

Byrne and Belton[9] made an accurate determination of the rate constant $\Phi_f$ for reaction of $N_2$ with high purity iron and Fe–C alloys at 1550–1700°C, by measuring the rate of $^{15}N \rightarrow {}^{14}N$ isotope exchange that occurs on the iron surface, as represented by

$$\log \Phi_f = -\ \frac{6340 \pm (710)}{T}\ + 1.85 \ (\pm 0.38) \tag{2.20}$$

where the rate constant $\Phi_f$ is in units of g N cm$^{-2}$min$^{-1}$atm$^{-1}$. The apparent heat of activation $\Delta H^* = 121.4$ kJ mol$^{-1}$ is much lower than the value expected for the rate of dissociation of $N_2$. As pointed out by Byrne and Belton, the rate of chemisorption of $N_2$ is presumably controlling the reaction mechanism.

The surface active elements dissolved in iron, e.g. O, S, Se, Te, are known to lower the rate of nitrogen transfer across the iron surface. On the basis of the experimental rate data with liquid iron containing O and S, as given in various publications (Refs. 6–10) and the surface tension data, the effects of O and S on the fraction of vacant sites, $1-\theta$, in the chemisorbed layer may be represented by

$$1 - \theta = \ \frac{1}{1 + 260 \ (\%O + \%S/2)} \tag{2.21}$$

which is a slightly simplified form of the equation that was derived by Byrne and Belton.

For the chemical reaction-controlled nitrogen removal from liquid iron (or steel) in reduced pressures or in an inert gas stream with very low $N_2$ pressure

$$2[N] \rightarrow N_2(g) \tag{2.22}$$

the integrated form of the rate equation is

$$\frac{1}{\%N} - \frac{1}{\%N_o} = \frac{100A}{\rho V} \ \Phi_r \ (1 - \theta)t \tag{2.23}$$

where $\%N_o$ is the initial nitrogen content and $\Phi_r$ the rate constant $= \Phi_f/K$. The solubility of $N_2$ in liquid iron or low alloy steel is given by

$$\log K \ (= [\%N]^2/p_{N_2} \ (\text{atm}) = -\ \frac{376}{T} - 2.48 \tag{2.24}$$

Combining this with equation (2.20) gives for the rate constant $\Phi_r$ in gN cm$^{-2}$min$^{-1}$%N$^{-1}$

$$\log \Phi_r = -\ \frac{5964}{T}\ + 4.33 \tag{2.25}$$

*2.1.1b Rate of decarburisation of liquid Fe–C alloys by CO$_2$*

Sain and Belton[11] determined the rate of decarburisation of high purity Fe–C alloys by $CO_2$ at temperatures between 1160 and 1600°C, under conditions where mass transport of reactants was not rate determining. For the reaction

$$CO_2(g) + [C] \rightarrow 2CO(g) \tag{2.26}$$

the rate of decarburisation was found to be controlled by the rate of chemisorption of $CO_2$ on the melt surface. That is,

$$\frac{d[\%C]}{dt} = -\Phi_f(1-\theta)p_{CO_2} \tag{2.27}$$

for which the following rate constant $\Phi_f$, g C cm$^{-2}$min$^{-1}$atm$^{-1}$, was obtained as a function of temperature.

$$\log \Phi_f = -\frac{5080}{T} + 2.65 \tag{2.28}$$

The temperature coefficient corresponds to an apparent heat of activation $\Delta H^* = 97.3$ kJ mol$^{-1}$ which, as argued by Sain and Belton, is indicative of the rate of chemisorption being the controlling reaction mechanism. If the rate were controlled by the dissociation of $CO_2$ in the adsorbed layer, the apparent heat of activation would have been much higher, e.g. 375 to 425 kJ mol$^{-1}$.

The sulphur in iron retards the rate of decarburisation, i.e. rate of $CO_2$ chemisorption, in much the same way as it affects the rate of $N_2$ chemisorption. This is demonstrated by the experimental results of Sain and

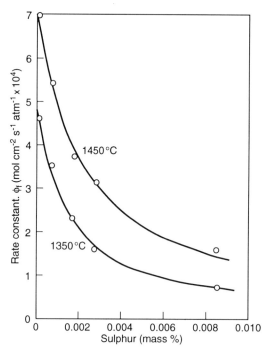

Fig. 2.3   Effect of sulphur on the rate constant for decarburisation of graphite-saturated liquid iron. From Ref. 11.

Belton reproduced in Fig. 2.3. The curves were calculated for the values of $(1-\theta)$, derived from the surface tension data on the Fe–C–S melts. A similar result is obtained by using equation (2.21) with due correction for the effect of carbon on the activity coefficient of sulphur $f_S^C$, thus for the Fe–C–S melts

$$(1 - \theta) = \frac{1}{1 + 130 f_S^C [\%S]} \tag{2.29}$$

where $\log f_S^C = 0.11 [\%C]$.

## 2.2 Fick's Diffusion Laws

### 2.2.1 First Law-Steady State Diffusion

The quantity of diffusing substance which passes per unit time through unit area of a plane perpendicular to the direction of diffusion, known as the flux $J$, is proportional to the concentration gradient of the diffusing substance

$$J = - D \frac{dC}{dx} \tag{2.30}$$

The coefficient $D$ is the diffusivity of the substance in the medium; $C$ is the concentration of the substance per unit volume and $x$ the distance in the direction of diffusion.

The steady state diffusion is illustrated graphically in Fig. 2.4 for gas diffusing through a permeable diaphram of thickness $\Delta x$.

### 2.2.2 Second Law–Nonsteady State Diffusion

The rate of accumulation of diffusing substance in a given volume element is the difference between the inward and outward flux. In other words, what goes in and does not come out, stays there.

The rate of concentration change – $dC/dt$ resulting from flux over a distance $dx$ is

$$J = - \frac{dC}{dt} dx$$

Hence, the change in flux with distance is

$$\frac{dJ}{dx} = - \frac{dC}{dt}$$

Invoking the first law and re-arranging gives the second law.

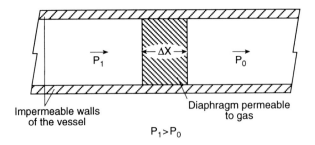

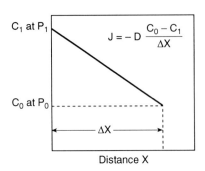

Fig. 2.4   Illustration of steady state diffusion.

$$\frac{dJ}{dx} = -\frac{d}{dx}\left(D\,\frac{dC}{dx}\right)$$

$$\frac{dC}{dt} = D\,\frac{d^2C}{dx^2} \tag{2.31}$$

The solution of this equation depends on the geometry and on the boundary conditions of the medium in which the dissolved substance is diffusing.

For the boundary conditions,
$C_o$ = initial uniform concentration
$C_s$ = constant surface concentration
$D$ = constant diffusivity

the solution of Fick's second law gives the relation in terms of dimensionless variables shown in Fig. 2.5 for the sphere and infinite cylinder of radius $L$, and slab of thickness $2L$. The fractional saturation, or desaturation is defined by

$$f = \frac{C_m - C_o}{C_s - C_o}$$

where $C_m$ is the mean concentration of the diffusate.

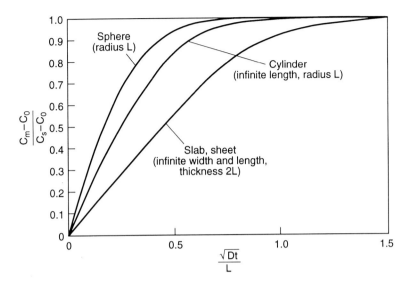

Fig. 2.5   Mean concentration or fractional saturation of slab, cylinder, or sphere of uniform initial concentration $C_0$ and constant surface concentration $C_s$, $C_m$ is the mean concentration at time $t$.

## 2.2.3 INTERDIFFUSION IN BINARY METALLIC SOLUTIONS

In a study of diffusion in alpha brass, Smigelskas and Kirkendall[12] found that zinc moves much faster than copper as evidenced by the relative motion of inert molybdenum wire placed on the sample as markers. This observation of the difference in the drift velocities of components of a solid solution, known as the 'Kirkendall effect,' led to numerous definitions of the coefficient of diffusion in binary metallic solutions, e.g. chemical diffusivity, intrinsic diffusivity or volume diffusivity. Subsequently, Stark[13] showed that the interdiffusivity $D$ is an invariant of binary diffusion motion as formulated below

$$D = - \frac{C_2 \bar{V}_2 J_1}{\partial C_1 / \partial x} - \frac{C_1 \bar{V}_1 J_2}{\partial C_2 / \partial x} \tag{2.32}$$

where $\bar{V}$s are the partial molar volumes. For dilute solutions, i.e. $C_2 \to 0$ and $C_1 \bar{V}_1 \to 1$, the above equation reduces to the form of Fick's first law

$$J_2 = - D \, \frac{\mathrm{d}C_2}{\mathrm{d}x}$$

Darken[14] and Hartley and Crank[15] independently derived the phenomenological equation for diffusion in a binary system 1–2, showing the thermodynamic effect on the interdiffusivity

$$D = (C_1 \bar{V}_1 D_2^* + C_2 \bar{V}_2 D_1^*) \; \frac{d \ln a_2}{d \ln C_2} \tag{2.33}$$

where $D_1^*$ and $D_2^*$ are the self-diffusivities of components 1 and 2. For dilute solutions, $C_2 \rightarrow 0$, $D = D_2^*$.

### 2.2.4 DIFFUSION IN IONIC MEDIA (SLAGS)

Polymeric melts, i.e. molten slags and glasses, are ionic in nature consisting of positively charged ions, known as cations, and negatively charged complex silicate, aluminate and phosphate ions, known as anions. Cations are the primary mobile species in polymeric melts. For the limiting case of $D_a^*$ (anion) $<< D_c^*$ (cation), the chemical diffusivity of the cation is about the same as the self-diffusivity, $D_c \sim D_c^*$.

## 2.3 MASS AND HEAT TRANSFER THROUGH REACTION PRODUCT LAYER

### 2.3.1 PARABOLIC RATE OF OXIDATION OF IRON

The scale forming in the oxidation of iron, or low alloy steels, consists of three layers of oxides: wustite on the iron surface followed by magnetite then haematite on the outer surface of the scale. From several experimental studies it was found that the average wustite/magnetite/haematite thickness ratios were about 95:4:1. In iron oxides, the diffusivity of iron is greater than the oxygen diffusivity. The rate of oxidation is controlled primarily by diffusion of iron through the wustite layer from the iron–wustite interface to the wustite–gas or wustite–magnetite interface, then through the thin layers of magnetite and haematite.

The flux of iron through the wustite layer is by Fick's first law

$$J_{Fe} = D \; \frac{dC}{dX}$$

where $D$ is the iron diffusivity, $C$ the concentration of Fe in wustite and $X$ the thickness of the wustite layer. Upon integration

$$J_{Fe} X = D(C' - C'') \tag{2.34}$$

where $C'$ and $C''$ are the iron concentrations in the oxide at the iron–wustite and wustite–gas (or magnetite) interfaces respectively. The rate of increase in scale thickness with the flux of iron is

$$J_{Fe} = \bar{C} \; \frac{dX}{dt} \tag{2.35}$$

where $\bar{C}$ is the average iron content of wustite. From equations (2.34) and (2.35)

$$\bar{C} \, \frac{X dX}{dt} = D \, (C' - C'')$$

which upon integration gives the parabolic rate equation,

$$X^2 = \frac{2D}{\bar{C}} \, (C' - C'') t \qquad (2.36)$$

or
$$X^2 = \lambda \, t \qquad (2.37)$$

where $\lambda$ is the parabolic rate constant, cm$^2$ (scale) s$^{-1}$

If the measurement of the rate of oxidation is made by the thermo-gravimetric method, then the parabolic rate constant $k_p$ would be in units of (gO)$^2$cm$^{-4}$s$^{-1}$. From the compositions and densities of the oxides, with the relative thickness ratios of wustite:magnetite:haematite = 95:4:1, the values of $k_p$ and $\lambda$ are related as follows.

$$k_p, \, \text{(gO)}^2 \text{cm}^{-4} \text{s}^{-1} = 1.877\lambda, \, \text{cm}^2 \, \text{(scale) s}^{-1} \qquad (2.38)$$

Many experimental studies have been made of the rate of oxidation of iron in air and oxygen at temperatures 600 to 1300°C. The temperature dependence of the parabolic rate constant is shown in Fig. 2.6; the references to previous studies denoted by different symbols are given in a paper by Sheasby *et al.*[16]

From theoretical considerations, Wagner[17] derived the following equation for the parabolic rate constant in terms of the activity of oxygen and self-diffusivities of the mobile species in the scale. For the case of wustite with $D^*_{Fe} \gg D^*_{O}$, Wagner equation is simplified to

$$\lambda = 2 \int_{a'_O}^{a''_O} \frac{N_O}{N_{Fe}} \, D^*_{Fe} d \, (\ln a_O) \qquad (2.39)$$

where $a'_O$ and $a''_O$ are oxygen activities at the iron–wustite and wustite–magnetite interfaces, respectively.

The values of $k_p = 1.877\lambda$ calculated from the measured tracer diffusivity of iron[18,19] using Wagner's equation, are consistent with the average experimental values in Fig. 2.6, which is represented by the following equation.

$$\log k_p = -\frac{8868}{T} + 0.9777 \qquad (2.40)$$

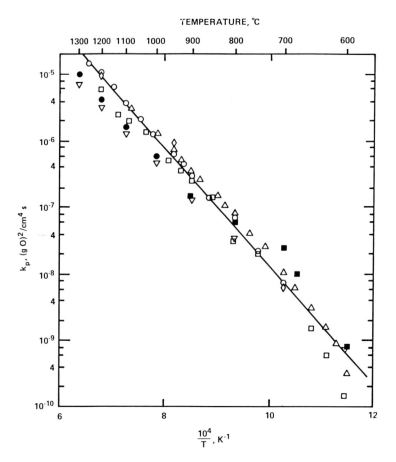

Fig. 2.6    Temperature dependence of the parabolic rate constant for oxidation of iron in air or oxygen with or without $H_2O$. From Ref. 16.

### 2.3.2 HEAT-TRANSFER CONTROLLED CALCINATION OF LIMESTONE

In the rate measurements made by Turkdogan *et al.*[20] on the calcination of relatively large spheroidal particles of limestone (3 to 14 cm dia), the centre and surface temperatures of the samples were measured with the Pt/Pt–Rh thermocouples during calcination. As noted from an example of the experimental data in Fig. 2.7, after about 50 percent calcination, with the formation of an outer casing of CaO, the centre temperature of the $CaCO_3$ core remained constant at 897°C until all the carbonate had decomposed.

The dissociation pressures of reagent gradient $CaCO_3$ measured by Baker[21] and Hills[22] are represented by the following equation.

$$\log p_{CO_2} = -\frac{8427}{T} + 7.169 \qquad (2.41)$$

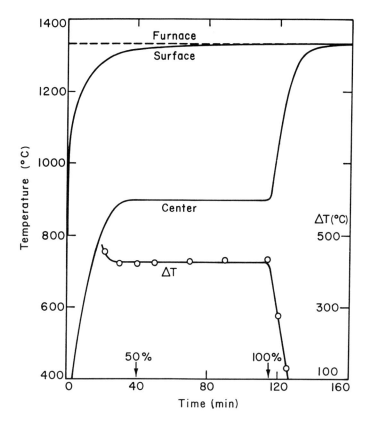

Fig. 2.7   Variation of surface and centre temperatures during calcination of 8.3 cm diameter limestone sphere in argon at 1 atm. From Ref. 20.

In the calcination experiments in argon at 1 atm pressure, the $CO_2$ pressure will be about 1 atm at the calcination front within the particles, for which the calculated centre temperature would be 902°C. This dissociation temperature is very close to the measured centre temperature 897°C, in the calcination of a high grade natural limestone, containing 2 mol% $MgCO_3$.

On the basis of these experimental findings, the rate controlling step was considered to be heat transfer through the porous CaO outer casing around the limestone core of the spheroidal sample for which the following rate equation would apply.

$$[3 - 2F - 3\,(1 - F)^{2/3}] = 6\kappa_e \left( \frac{T_s - T_i}{\rho \Delta H r_o^2} \right) t + C \qquad (2.42)$$

where $F$ is the fraction of $CO_2$ removed in calcination time t, $\kappa_e$ the effective thermal conductivity of burnt lime, $T_s$ the outer surface temperature, $T_i$ that at the centre, $\rho$ the bulk density of the limestone in units of mol $cm^{-3}$, $\Delta H$ the enthalpy of calcination, $r_o$ the initial radius of the spheroidal limestone particle and $C$ is a constant (a negative number) that takes account of all early time departures from the assumed boundary conditions for which equation (2.42) applies.

For the period of calcination when the temperature difference $\Delta T = T_s - T_i$ remains essentially unchanged, the effective thermal conductivity of burnt lime can be determined from the weight loss data using equation (2.42). Thus, with the notation $[3-2F-3(1-F)^{2/3}] = Y$,

$$\kappa_e = \frac{dY}{dt}\left(\frac{\rho\Delta H r_o^2}{6\Delta T}\right) \tag{2.43}$$

An example of the calcination rate data is given in Fig. 2.8 for a spheroidal 14 cm diameter (3.58 kg) limestone. Beyond 50 percent calcination, the measurements are in accord with the rate equation (2.42). The linear variation of $(dY/dt)r^2_o$ with the temperature difference $\Delta T$ across the lime layer, shown in Fig. 2.9, is in accord with the equation (2.43). The slope of the line with the appropriate values of $\Delta H$ and $\rho^*$, gives for the effective thermal conductivity $\kappa_e = 0.0053$ J cm$^{-1}$s$^{-1}$K$^{-1}$ for burnt lime with a porosity of about 50%.

## 2.4 REACTION OF GASES WITH POROUS MATERIALS

The rate of a heterogeneous reaction between a fluid and a porous medium is much affected by the counter-diffusive flow of fluid reactants and products through pores of the medium. Complexities of the rates of reaction of gases with porous materials are highlighted in this section with examples on the oxidation of carbon and the reduction of iron oxide which are relevant to some of the pyrometallurgical processes related to iron and steelmaking.

### 2.4.1 MATHEMATICAL ANALYSIS OF THIELE (1939)

The first mathematical analysis of a dual reaction rate-diffusion controlled process on the pore surface is attributed to the work of Thiele.[23] In order to obtain an analytical expression in a closed form for the rate of reaction on the pore surface, Thiele considered an ideal simple pore structure with the following assumptions.

1. The reactant molecules of the fluid are transported into and within the porous structure of a granule by a diffusive flux through a concentration gradient.
2. There is no volume change in the fluid medium resulting from the reaction on the pore surface, i.e. no pressure buildup in pores.

*For the limestone containing 2 mol % $MgCO_3$, $\Delta H$ is about 159.7 kJ mol$^{-1}$ and the carbonate density $\rho$ = 0.0248 mol cm$^{-3}$.

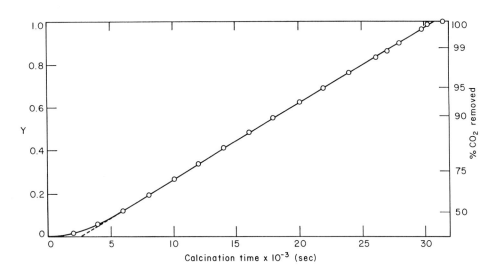

Fig. 2.8   Calcination rate of 14 cm diameter (3.58 kg) limestone spheroid in air at 1135°C, plotted in accord with equation 2.42. From Ref. 20.

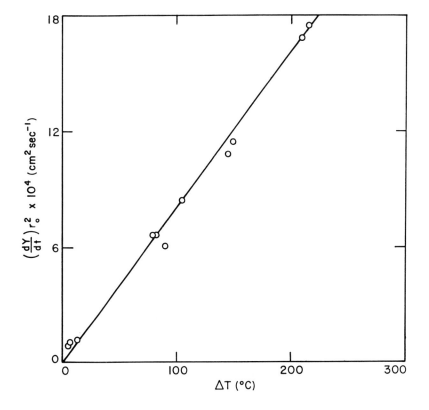

Fig. 2.9   Variation of the rate of calcination, $(dY/dt)r_o^2$, with the temperature difference across the calcinated layer. From Ref. 20.

3. The surface diffusion in the adsorbed layer on the pore wall is negligible.
4. The pores are interconnected, uniform in cross sections and the pore surface area remains unchanged.
5. The reverse reaction at the pore wall is negligible and the steady state conditions prevail.
6. The temperature across the particle remains uniform.

In a detailed study of the reaction rates on catalyst pores, Wheeler[24] and Weisz and Prater[25] made an extensive use of Thiele's analysis in deriving rate equations for reactions on pore surfaces. For example, for a first order type reaction involving a gaseous species $i$, the rate of reaction per spherical porous pellet of radius $r$ is given by the following expression.

Rate per pellet (mol s$^{-1}$) =

$$4\pi r^2 C_i \,(\Phi \rho S D_e)^{1/2} \left\{ \frac{1}{\tanh\,(3\psi)} - \frac{1}{3\psi} \right\} \qquad (2.44)$$

where $C_i$ = molar concentration per cm$^3$ of reactant in the gas stream outside the pellet,

$\Phi$ = specific rate constant of chemical reaction per unit area of the pore wall,

$\rho$ = molar bulk density of the porous medium,

$S$ = internal pore surface area per unit mass, cm$^2$mol$^{-1}$,

$D_e$ = effective gas diffusivity, characteristic of the porous medium, cm$^2$s$^{-1}$,

$\psi$ = Thiele's dimensionless parameter defined by

$$\psi = \frac{r}{3} \left( \frac{\Phi \rho S}{D_e} \right)^{1/2} \qquad (2.45)$$

Three rate limiting cases are given below.

(a) When $\psi$ is small, e.g. $\psi<0.2$, with (i) decreasing particle size, (ii) decreasing temperature, i.e. decreasing $\Phi/D_e$, and (iii) decreasing pressure, i.e. increasing $D_e$, the dimensionless parameters in parenthesis in equation (2.44) approach the value of $\psi$ in equation (2.45) and equation (2.44) is simplified to

$$\text{Rate per pellet (mol s}^{-1}) = \frac{4}{3}\,\pi r^3 \Phi \rho S C_i \qquad (2.46)$$

This is the limiting case for almost complete pore diffusion, hence internal burning.

(b) When $\psi$ is large, e.g. $\psi>2$, with (i) increasing particle size, (ii) increasing temperature, i.e. increasing $\Phi/D_e$ and (iii) increasing pressure, i.e. decreasing $D_e$, $\tanh\,(\psi) \rightarrow 1$ and equation (2.44) is simplified to

$$\text{Rate per pellet (mol s}^{-1}) = 4\pi r^2 (\Phi \rho S D_e)^{1/2} C_i \qquad (2.47)$$

The diffusion-controlled reaction at this limiting case is confined to pore mouths at the outer surface of the pellet, and hence the rate per pellet is proportional to the external geometrical surface area of the pellet. Furthermore, since $D_e$ does not vary much with temperature, the apparent heat of activation for the reaction at this limiting case is about one-half of that for the case (a) of almost complete pore diffusion.

(c) With large particles, high temperatures and low velocity gas flows, the rate is ultimately controlled by mass transfer in the gas-film boundary layer, for which the rate per pellet would be proportional to the particle diameter.

### 2.4.2 OXIDATION OF CARBON IN $CO_2$–CO MIXTURES

Most forms of carbon are porous, therefore the rate of oxidation, i.e. gasification of carbon, is decisively affected by the pore structure, internal pore surface area, effective gas diffusivity in the pores and particle size as outlined in the previous section.

In a critical review of the oxidation of carbon, Walker et al.[26] gave a detailed and a comprehensive account of (up to 1959) experimental and theoretical studies on this subject. In the late 1960s, Turkdogan and co-workers[27–29] pursued further experimental work on the oxidation of several types of carbon in $CO_2$–CO mixtures at temperatures up to 1300°C and at pressures of 0.01 to 10 atm. A few salient features of their findings are briefly given here.

The effect of particle size on the rate of oxidation of high purity graphite spheres in $CO_2$ at atmospheric pressure and temperatures 900, 1000 and 1100°C is shown in Fig. 2.10. In accord with Thiele analysis, for smaller particles the rate per pellet is proportional to the cube of particle diameter, indicating chemical reaction control in essentially uniform internal burning. With larger particle size, the rate of oxidation is controlled by limited gas diffusion into pores, hence the rate is proportional to the square of particle diameter in accord with equation (2.47). Then, there is the mixed control for intermediate particle sizes.

In rate measurements with many small granules, the rate is given in terms of the fraction $F$ of mass oxidised, for which the integrated forms of the rate equations (2.46) and (2.47) are as follows.

a. For $\psi < 0.2$ complete internal burning:

$$\ln(1 - F) = -\Phi S C_i t \qquad (2.48)$$

b. For $\psi > 2$ external burning:

$$1 - (1 - F)^{1/3} = \frac{(\Phi \rho S D_e)^{1/2}}{\rho r_o} C_i t \qquad (2.49)$$

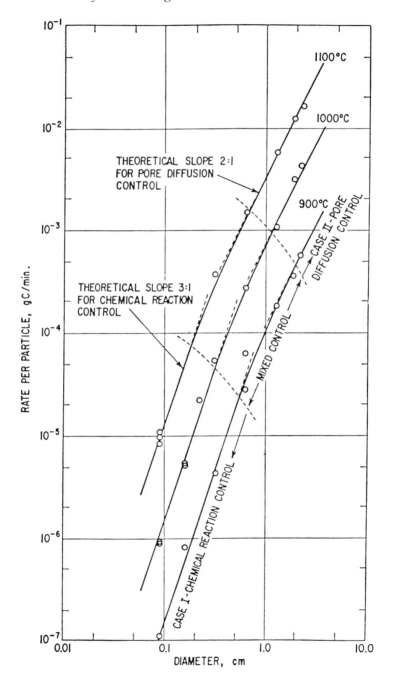

Fig. 2.10   Rate of oxidation of electrode graphite as a function of particle diameter for 0.96 atm $CO_2$. From Ref. 27.

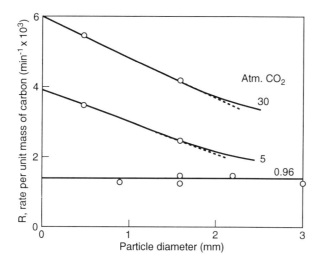

Fig. 2.11   Effect of particle size on the rate of oxidation of electrode graphite in pure carbon dioxide at 1000°C. From Ref. 28.

In the integration of equation (2.47) to give the above equation, due account is taken of the shrinking size of the spherical particle of initial radius $r_o$ during gasification.

As is seen from the rate data in Fig. 2.11, an increase in $CO_2$ pressure increases the particle size effect in lowering the extent of internal oxidation. In the oxidation of various types of carbon in pure $CO_2$ at pressures of 0.01 to 10 atm, the rate was found to be proportional to the square root of the $CO_2$ pressure. In the present re-assessment of these rate data, the author is now of the opinion that this apparent square root relationship is purely empirical and not to be attributed to some hypothetical reaction mechanism.

On the assumption that both CO and $CO_2$ are chemisorbed on the pore walls of the carbon and the rate of oxidation is due to the rate of dissociation of chemisorbed $CO_2$, for the limiting case of complete internal burning the rate equation will be in the following form

$$\text{Rate} = \frac{\Phi' p_{CO_2}}{1 + k_1 p_{CO} + k_2 p_{CO_2}} \ t \qquad (2.50)$$

where $\Phi'$ is the rate constant ($\text{min}^{-1}\text{atm}^{-1}CO_2$) for a given temperature and a particular type of carbon; the constants $k_1$ and $k_2$ are associated with the chemisorption of CO and $CO_2$. It should be noted that in compliance with the notation in equation (1.69) for the ideal monolayer, the constants $k_1$ and $k_2$ are reciprocals of the respective activity coefficients of the adsorbed species, i.e. $k_1 = 1/\varphi_{CO}$ and $k_2 = 1/\varphi_{CO_2}$; the fraction of vacant sites being $(1 - \theta) = (1 - \theta_{CO} - \theta_{CO_2})$.

Examples of the experimental data on the initial rate of oxidation of granular ($\approx 0.5$ mm dia.) electrode graphite and metallurgical coke are given in Figs. 2.12 and 2.13, reproduced from a previous publication.[28] Upon re-assessment of these experimental data, the following equations are obtained for the temperature dependence of $\Phi'$, $\varphi_{CO} = 1/k_1$ and $\varphi_{CO_2} = 1/k_2$.

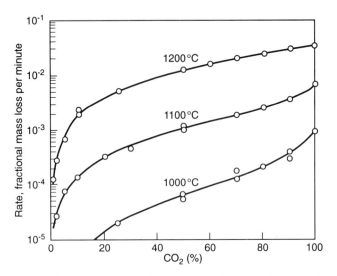

Fig. 2.12   Effects of temperature and gas composition on the rate of oxidation of electrode graphite granules (~0.5 mm dia.) in $CO_2$–$CO$ mixtures at 0.96 atm total pressure. From Ref. 28.

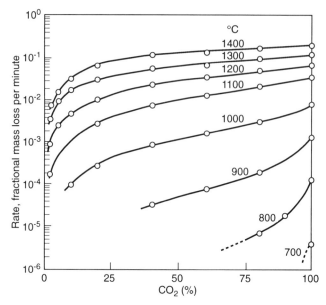

Fig. 2.13   Effect of temperature and gas composition on the rate of oxidation of granular metallurgical coke (~0.5 mm dia.) in $CO_2$–$CO$ mixtures at 0.96 atm total pressure. From Ref. 28.

For electrode graphite:

$$\log \Phi' \ (\text{min}^{-1}\text{atm}^{-1}\text{CO}_2) = - \ \frac{16{,}540}{T} + 10.75 \tag{2.51}$$

$$\log \varphi_{\text{CO}} \ (\text{atm}) = - \ \frac{8719}{T} + 4.84 \tag{2.52a}$$

$$\log \varphi_{\text{CO}_2} \ (\text{atm}) = - \ \frac{590}{T} - 0.072 \tag{2.52b}$$

For metallurgical coke:

$$\log \Phi' \ (\text{min}^{-1}\text{atm}^{-1}\text{CO}_2) = - \ \frac{16{,}540}{T} + 11.37 \tag{2.53}$$

$$\log \varphi_{\text{CO}} \ (\text{atm}) = - \ \frac{2117}{T} + 0.27 \tag{2.54a}$$

$$\log \varphi_{\text{CO}_2} \ (\text{atm}) = \frac{3840}{T} - 3.45 \tag{2.54b}$$

It should be noted that these equations fit the measured initial rates of oxidation within a factor of about 1.5 for temperatures 800 to 1200°C and for $CO_2$ pressures 0.03 to 3 atm. For easy comparison of the reactivities of electrode graphite and metallurgical coke, the numerical values are given below from the above equations for 900 and 1200°C.

| Temperature °C | Electrode graphite | | | Metallurgical coke | | |
|---|---|---|---|---|---|---|
| | $\Phi'$ | $\varphi_{\text{CO}}$ | $\varphi_{\text{CO}_2}$ | $\Phi'$ | $\varphi_{\text{CO}}$ | $\varphi_{\text{CO}_2}$ |
| 900 | $4.5 \times 10^{-4}$ | 0.0025 | 0.27 | $1.9 \times 10^{-3}$ | 0.029 | 0.67 |
| 1200 | $3.3 \times 10^{-1}$ | 0.083 | 0.34 | $1.4 \times 10^{0}$ | 0.068 | 0.14 |

After about 3 to 5 percent of initial oxidation, the pore surface area of the coke samples were found to be four to five times greater than the graphite samples; this is consistent with the $\Phi'$ values for coke being greater than for graphite by a similar factor. The extent of CO adsorption on the pore walls of electrode graphite or coke is greater than the $CO_2$ adsorption.

There are variations in the reported values of the apparent heat of activation for oxidation of the electrode (or reactor grade) graphite in $CO_2$. For example, Gulbransen et al.[30] obtained $\Delta H^* = 368$ kJ mol$^{-1}$ while according to Blackwood's work[31] $\Delta H^* = 260$ kJ mol$^{-1}$; in the present case $\Delta H^* = 317$ kJ mol$^{-1}$. Reference should be made also to the papers of Ergun,[32] Hedden and Löwe,[33] and by Grabke[34] for various other interpretations of the kinetics of oxidation of carbons.

Aderibigbe and Szekely[35] also investigated the oxidation of metallurgical coke in $CO_2$–CO mixtures at 850 to 1000°C. Their rate constants $K_1(\equiv \Phi')$ are similar to those given by equation (2.53); their estimate of the apparent heat of activation being $\Delta H^* = 249 \pm 47$ kJ mol$^{-1}$.

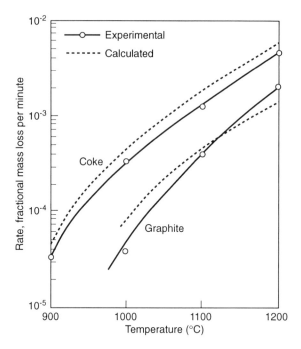

Fig. 2.14   Effect of temperature on rate of oxidation of coke and graphite (1.9 and 2.2 cm dia. spheres, respectively) in 50:50 $CO_2$–CO at 1 atm. From Ref. 36.

A mathematical analysis was made by Tien and Turkdogan[36] to formulate the pore diffusion effect on partial internal burning of relatively large carbon particles. The experimentally determined initial rates of oxidation of coke and graphite spheres ($\approx 2$ cm dia.) in 1:1 $CO_2$–CO mixture at 1 atm are compared in Fig. 2.14 with the values calculated from the mathematical analysis. The results of such calculations are summarised in Fig. 2.15, showing critical particle size and temperature for the limiting rate controlling processes at 1 atm for 100% $CO_2$ and 1:1 $CO_2$–CO mixture. In each diagram, the lower curve is for 80 percent internal burning; therefore, in the region below this curve there is almost complete pore diffusion. The upper curve is for 20 percent internal burning; therefore, in the region above this curve, the pore diffusion control predominates.

### 2.4.2a ELECTRODE CONSUMPTION IN EAF STEELMAKING

The foregoing experimental data on the rate of oxidation of graphite will give only a partial estimate of the rate of electrode consumption in the electric arc furnace (EAF).

The net consumption is the result of various modes of wear that the electrodes are subjected to in the electric furnace environment. It can be subdivided into longitudinal consumption, consisting of such variables as

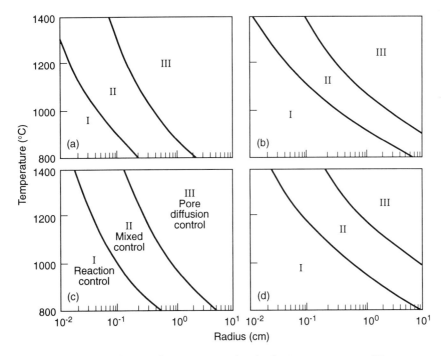

Fig. 2.15   Critical radius and temperature for the limiting rate controlling processes at 1 atm pressure. (a) Coke, 100 per cent $CO_2$; (b) Coke, 1:1 $CO:CO_2$; (c) Graphite, 100 per cent $CO_2$; (d) Graphite, 1:1 $CO:CO_2$. From Ref. 36.

arc vaporisation, butt losses, mechanical erosion due to interaction with the steel bath etc., and transverse consumption due to oxidation of the tapered sidewall of the electrode. It is generally believed that a taper which results in a tip diameter of about 70 percent of the original electrode diameter will produce low consumption rates. The net consumption is about 70 to 90 percent of the total consumption, the remaining 10 to 30 percent consumption being due to breakage.

Reference may be made to a paper by Jung *et al.*[36a] for a detailed discourse on electrode wear in the EAF steelmaking.

## 2.4.3 REDUCTION OF IRON OXIDES

In view of its practical importance to the understanding and control of ironmaking processes, a great deal of research has been done on the gaseous reduction of iron oxides and iron ores. Because of the porous nature of iron oxides and the reduction products, the interpretation of the reduction rate data is inherently complex.

The formation of product layers during the gaseous reduction of dense sintered haematite and magnetite pellets or natural dense iron ore particles is a well-known phenomenon, as shown in Fig. 2.16. In several studies

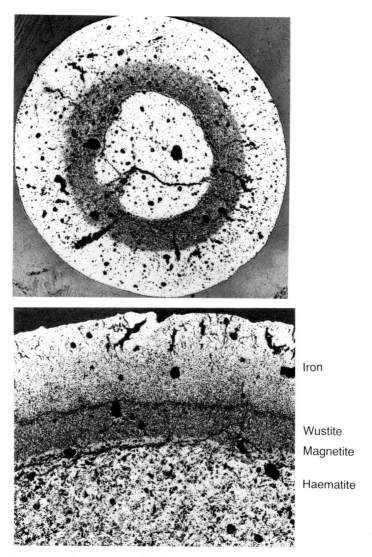

Fig. 2.16   Formation of reaction product layers in the gaseous reduction of haematite pellets.

made in the early 1960s[37–39] it was found that the thickness of the reduced iron layer, encasing the iron oxide core of the pellet, increased linearly with the reduction time. The measured rates were interpreted in terms of the rate-conrolling chemical reaction at the iron wustite interface; the diffusive fluxes of gases through the porous layers were assumed to be relatively fast. On the other hand, Warner[40] and Spitzer *et al.*[41] have expressed the view that the rate of gaseous reduction is much affected by the gaseous diffusional processes, e.g. the gas–film resistance at the pellet surface and particularly the resistance to diffusion in the porous product layers. The rate measurements made by Turkdogan and Vinters[42] on the reduction of haematite pellets in $H_2$–$H_2O$ and $CO$–$CO_2$ mixtures have clearly demonstrated that the rate-controlling effects of gas diffusion into the pores of the oxide granules or pellets and through the porous iron layer dominate the reaction kinetics.

The sketch in Fig. 2.17 demonstrates three limiting rate controlling processes as outlined below.

(a) With fine granules there is internal reduction producing rosettes or platelets of metallic iron within the oxide particle; the rate in terms of mass fraction reduced, $F$, is independent of particle size.

(b) With large and dense oxide particles and at high temperatures, the gas diffusion is slow and the reaction is confined essentially to pore mouths on the outer surface of the particle, bringing about the development of a porous iron layer around the pellet. Because of the layer formation, this mode of reduction is often called 'topochemical reduction.' In the early stages of reduction, the porous iron layer is sufficiently thin for rapid gas diffusion, therefore the initial rate of reduction is controlled jointly by (i) gas diffusion into the pore mouths of the oxide and (ii) reaction on the pore walls of the wustite. That is, with Thiele parameter $\psi > 2$, the initial rate of reduction is represented by the following proportionality for a given temperature and gas composition

$$\frac{dF}{dt} \propto \frac{\sqrt{\Phi SD_e}}{r} \qquad (2.55)$$

(c) When the porous iron layer becomes sufficiently thick, the rate of reduction will be controlled essentially by the counter current gas diffu-

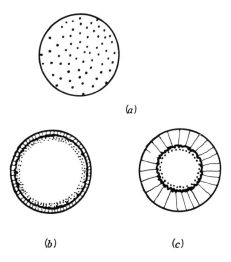

(a)

(b)                    (c)

Fig. 2.17  Schematic representation of (a) uniform internal reduction of wustite to iron, and (b), (c) topochemical reduction; (b) limiting mixed control (partial internal reduction) and (c) diffusion in porous iron as the rate-controlling step.

sion ($H_2$–$H_2O$ or $CO$–$CO_2$) for which the limiting rate equation, for a given temperature is as follows.

$$[3 - 2F - 3\,(1 - F)^{2/3}] = Y = \frac{D_e}{\rho r^2}\;\left(\frac{p_i - (p_i)_{eq}}{RT}\right)\;t + C \qquad (2.56)$$

where $p_i$ is the $H_2$ or $CO$ partial pressure in the gas stream and $(p_i)_{eq}$ that for the iron–wustite equilibrium and $C$ a constant (a negative number) that takes account of all early time departures from the assumed boundary conditions for this limiting case.

(d) At high reduction temperatures, with large oxide pellets and low velocity gas flows, the rate of reduction is controlled primarily by mass transfer in the gas-film layer at the pellet surface. In this limiting case the rate is inversely proportional to the square of the particle diameter.

The experimental data are given in Fig. 2.18 showing the particle size effect on the initial rate of reduction of haematite granules or pellets in

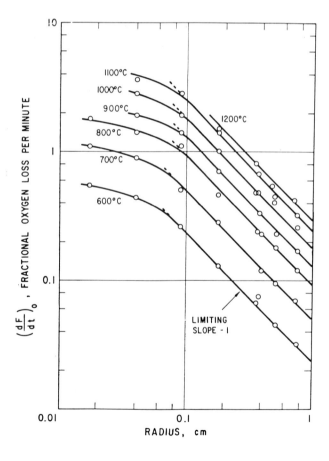

Fig. 2.18   Initial rate in hydrogen at 0.96 atm. as a function of particle radius at indicated temperatures. From Ref. 42.

atmospheric pressure of hydrogen. It is seen that for pellets of radius more than 1 mm, the reduction is confined to the outer surface in accord with the equation (2.55).

For uniform internal reduction, the particle diameter should be less than 0.1 mm. The effect of particle size on the rate of reduction of haematite granules, d ln $(1 - F)/dt$, is shown in Fig. 2.19. Small extrapolation to the hypothetical zero particle size gives the rate for uniform internal reduction

$$\ln (1 - F) = - \Phi S \left\{ p_i - (p_i)_{eq} \right\} t \tag{2.57}$$

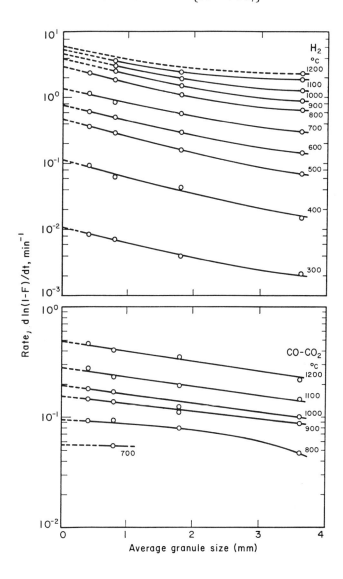

Fig. 2.19   Effect of granule size on the rate of internal reduction of haematite ore in $H_2$ and 90% CO + 10% $CO_2$ mixture. From Ref. 42.

As discussed in section 2.4.4, the pore surface areas of reduced iron oxides have been measured. These $S$ values in conjunction with the apparent rate constants $\Phi S$ give the specific rate constants for gaseous reduction of wustite. The temperature dependence of the specific rate constants $\Phi(H_2)$ and $\Phi(CO)$, derived from the reduction data, are given below.

$$\log \Phi (H_2) = - \frac{6516}{T} - 1.31 \tag{2.58}$$

$$\log \Phi (CO) = - \frac{7634}{T} - 1.80 \tag{2.59}$$

The ratio of the rate constants $\Phi(H_2)/\Phi(CO)$ is about 20 at 1200°C and increases to 50 at 700°C.

When the thickness of the reduced porous iron layer exceeds 1 mm, the subsequent rate of reduction is controlled primarily by gas diffusion through the porous iron layer. The reduction data plotted in accord with equation (2.56) usually give elongated S-shaped curves as in Fig. 2.20 for 15 mm diameter spheroidal haematite ore reduced in $H_2$ at 1 atm. From about 50% to 95% or 99% reduction, data are well represented by straight lines. The effective $H_2$–$H_2O$ diffusivities in the pores of the iron layer are derived from the slopes of the lines; details are given in the next section.

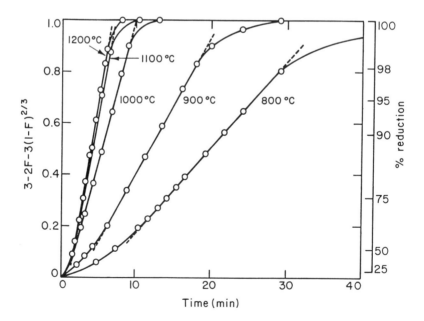

Fig. 2.20 Diffusion plot of reduction data for 15 mm diameter spheroidal haematite ore. From Ref. 42.

2.4.4 EFFECT OF REACTION TEMPERATURE ON PORE STRUCTURE OF REACTED
MATERIAL

(a) Reduced iron oxide:
The reaction temperature has a pronounced effect on the pore structure of
the reaction product as in calcination and reduction. This is well demonstrated by the SEM micrographs in Fig. 2.21.

The top left micrograph shows the fracture surface of dense haematite
ore which consists of agglomerates of spheroidal ore grains 2 – 5 μm
diameter. The other micrographs show the fracture surfaces of the reduced porous iron. The higher the reduction temperature, the coarser the
pore structure.

The effect of reaction temperature on the pore surface area of iron and
wustite formed by the reduction of haematite is shown in Fig. 2.22. The
higher the reduction temperature, the smaller the internal pore surface
area, i.e. the coarser the pore structure. The iron oxides reduced at temperatures below 800°C are known to be pyrophoric, which is a consequence of a fine pore structure with a large pore surface area as noted
from the data in Fig. 2.22.

The pore surface area of the iron formed by gaseous reduction depends
on the initial pore surface area of the iron oxide. This is demonstrated in
Fig. 2.23; for a given type of iron oxide, the pore surface area of iron
formed upon reduction increases with increasing initial pore surface area
of the oxide. However, there is a considerable overlap. For example, a
sintered oxide pellet with $S = 0.1$ m$^2$g$^{-1}$ and a haematite ore with $S = 10$
m$^2$g$^{-1}$, when reduced, both give porous iron with $S \approx 1$ m$^2$g$^{-1}$. That is the
pore surface area of the reduction or calcination product can be larger or
smaller than the pore surface area of the unreacted material. It is for these
reasons that care should be taken in comparing the rates of gaseous reduction of different types of metal oxides.

It is seen from the data in Fig. 2.24 that the effective diffusivities for H$_2$–
H$_2$O measured directly or calculated from the pore structure considerations,[43] agree well with those derived from the rate measurements within
the regime of pore-diffusion control, i.e. from the slopes of the lines in Fig.
2.20. The temperature has a small effect on gas diffusivities. A marked
effect of temperature on the effective diffusivity seen in Fig. 2.24 is due
mainly to the coarseness of the pore structure of iron when the oxide is
reduced at higher temperatures.

(b) Calcined limestone:
The effects of time and temperature on the pore surface area of burnt lime
are shown in Fig. 2.25; the initial pore surface area of the limestone used in
the calcination experiments was 0.07 m$^2$g$^{-1}$. The soft burnt lime, i.e. low
calcination temperature, being more reactive than the hard burnt lime, i.e.

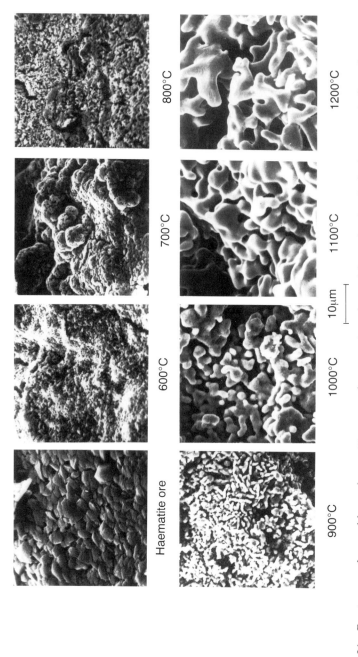

Fig. 2.21　Fracture surfaces of lump haematite ore and porous iron formed by reduction in hydrogen at indicated temperatures as viewed in the scanning electron microscope. From Ref. 42.

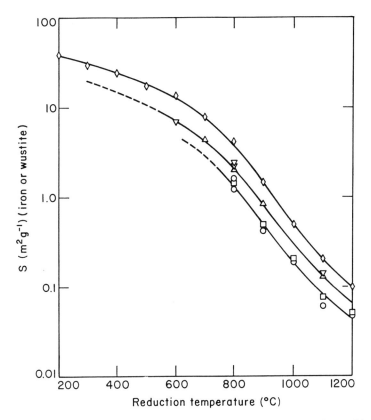

Fig. 2.22   Connected internal pore surface area of iron and wustite formed by reduction of haematite ores *A* and *B*; (a) ore *A* reduced to iron in $H_2$ ($\diamond$); (b) ore *B* reduced to iron in $H_2$ ($\triangledown$) and reduced to wustite in an $H_2$–$H_2O$ mixture ($\triangle$); (c) ore *B* reduced to iron in a $CO$–$CO_2$ mixture ($\square$) and reduced to wustite in another $CO$–$CO_2$ mixture ($\circ$). From Ref. 42.

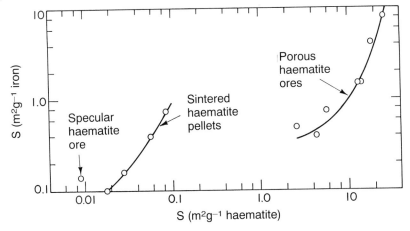

Fig. 2.23   Relations between the internal pore surface area of $H_2$ – reduced iron at 800°C and the pore surface area of the corresponding haematite ores or sintered pellets. From Ref. 42.

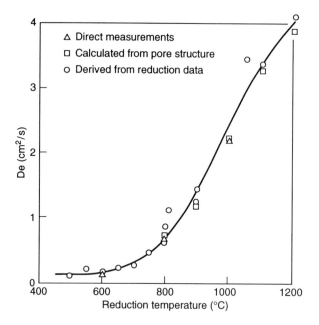

Fig. 2.24 $H_2$–$H_2O$ effective diffusivity derived from data for rate of reduction of haematite to iron is compared with that obtained by direct measurements and calculated from pore structure. From Ref. 43.

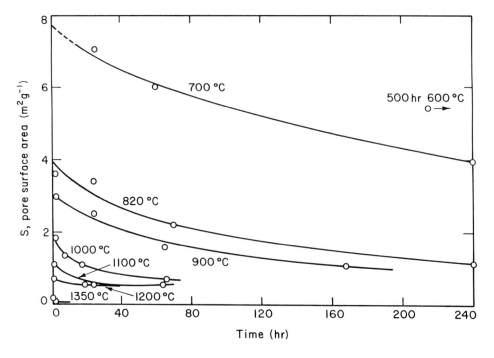

Fig. 2.25 Effect of heat-treatment time and temperature on the pore surface area of burnt lime. From Ref. 20.

**Table 2.1**   Measured CO–CO$_2$ effective diffusivities for electrode graphite.

| Temperature °C | % oxidised | $D_e$ (CO–CO$_2$) cm$^2$s$^{-1}$ |
|:---:|:---:|:---:|
| 500 | 0 | 0.006 |
| 500 | 3.8 | 0.016 |
| 500 | 7.8 | 0.020 |
| 500 | 14.6 | 0.045 |
| 700 | 0 | 0.009 |
| 700 | 3.8 | 0.023 |
| 700 | 7.8 | 0.029 |
| 700 | 14.6 | 0.067 |
| 800 | 0 | 0.012 |
| 900 | 0 | 0.014 |

high calcination temperature, is self-evident from the time and temperature effect on the pore surface area.
(c) Partially gasified carbon:
In the case of oxidation of carbons, the pore surface area increases during the
early stages of gasification then remains essentially unchanged. For example,
the pore surface area of electrode graphite with initial $S = 1$ m$^2$g$^{-1}$ increases to
about 2.5 m$^2$g$^{-1}$ after 10% gasification with only a small increase during
further gasification. For the metallurgical coke with initial $S = 4$ m$^2$g$^{-1}$, the
area increases to about 20 m$^2$g$^{-1}$ after 10% gasification then remains essentially unchanged during the subsequent stages of gasification.[29]

In the case of internal burning of carbon, the pore structure becomes
coarser as the gasification progresses, resulting in an increase in the effective gas diffusivity. A few examples of the measured CO–CO$_2$ effective
diffusivities for electrode graphite are given in Table 2.1, reproduced from
Ref. 29.

## 2.5 Vaporisation in Reactive Gases

There are two types of mechanisms of enhanced vaporisation in reactive
gases: (i) a chemical process involving the formation of volatile species,
e.g. oxides, sulphides, halides, carbonyls and hydroxides, and (ii) a transport process in which the vapour reacts with the gas diffusing toward the
vaporising surface.

The concept of diffusion-limited enhanced vaporisation, introduced by
Turkdogan *et al.*[44] is illustrated schematically in Fig. 2.26, using as an
example the vaporisation of liquid iron in a stream of an oxygen-inert gas
mixture. In this counter-current transport process, at some short distance
from the surface of the metal, the gaseous oxygen and metal vapour react
to form a metal oxide mist.

$$2M(g) + O_2(g) \rightarrow 2MO(s,l) \tag{2.60}$$

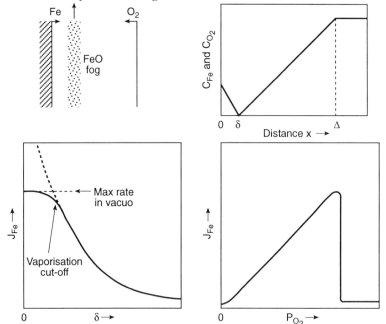

Fig. 2.26   Schematic representation of vaporisation of metals (iron as an example) in oxygen-bearing gas mixture flowing over the surface. From Ref. 44.

The formation of an oxide in the gas phase provides a sink for the metal vapour and oxygen resulting in a counterflux of these two gaseous species. Increasing the oxygen partial pressure in the gas mixture decreases the distance $\delta$ through which the metal vapour is diffusing, hence the rate of vaporisation increases. On further increasing the oxygen partial pressure, the flux of oxygen towards the surface of the metal eventually becomes greater than the equivalent counter-flux of the metal vapour. This results in the oxidation of the metal surface and cessation of vaporisation because the vapour pressure of the metal oxide is much lower than the vapour pressure of the metal. Just before this cutoff, the rate of vaporisation will be close to the maximum rate for free vaporisation *in vacuo*. This mechanism of enhanced vaporisation in an oxidising gas is the primary cause of the formation of metal oxide fumes in all pyrometallurgical processes. With the elimination of oxygen in air by burning natural gas and injecting steam in the vicinity of an exposed liquid steel (or hot metal) stream, the fume emission is drastically curtailed. In fact, it was on the basis of these findings that the fume suppression techniques have been implemented for many years in the U.S. Steel plants.

As is seen from some examples of the laboratory experimental results (Fig. 2.27) the rates of vaporisation of iron and copper increase with increasing partial pressure of oxygen until a maximum rate is obtained, beyond which vaporisation ceases. The results with various metals summarised in Fig. 2.28 show that the measured maximum rates of

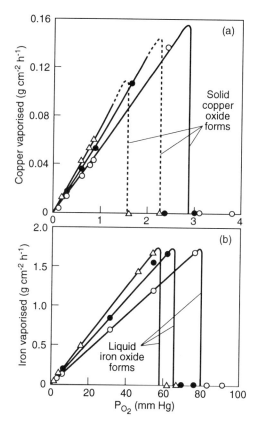

Fig. 2.27   Rates of vaporisation of copper at 1200°C and iron at 1600°C in argon–oxygen mixtures at 1 atm pressure as affected by gas velocity parallel to surface and partial pressure of oxygen. Gas velocity: (○) 40, (●) 60, and (△) 80 cm sec[-1]. From Ref. 44.

vaporisation agree well with those calculated for free vaporisation using equation (2.14). A quantitative analysis of these observations in terms of mass transfer is given in the paper cited.[44]

White fumes emitted from the blast furnace runner exposed to air, is another example of vaporisation-induced occurrence. The X-ray scan spectra of fume samples have revealed that the potassium and sulphur were the most dominant elements in the white fume samples. This finding can be accounted for by the following reaction mechanism.

The blast furnace slag contains 1.5 to 2.0% S and about 0.5% $K_2O$. Upon exposure to air, the potassium and sulphur will be evolved from the slag surface as potassium sulphate by the reaction

$$2(K^+) + (S^{2-}) + 2O_2 \rightarrow K_2SO_4(\text{white fume}) \qquad (2.61)$$

In addition, there is $SO_2$ evolution from the uncovered slag runner

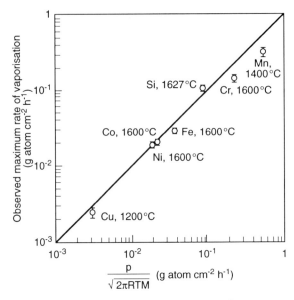

Fig. 2.28  Experimentally observed maximum rate of vaporisation of metals in argon–oxygen mixtures compared with rate of free vaporisation *in vacuo*. From Ref. 44.

$$(S^{2-}) + \frac{3}{2} O_2 \rightarrow (O^{-2}) + SO_2 \qquad (2.62)$$

In the oxidising atmosphere over the slag, exposed to air, the $SO_2$ is partly oxidised to $SO_3$. In the cooler ambient atmosphere away from the slag surface, $SO_3$ and $K_2SO_4$ will react with moisture in the air to form a complex sulphate $K_2SO_4.H_2SO_4$ which appears as a white fume. The X-ray scan spectra indicated an atom concentration ratio K/S of about one, in accord with the described reaction mechanism.

## 2.6 KINETICS OF GAS BUBBLES IN OXYGEN STEELMAKING

### 2.6.1 BUBBLE NUCLEATION AND GROWTH

The theory of homogeneous nucleation predicts a supersaturation of about $10^4$ atm for nucleation of gas bubbles in liquid metals with surface tension in the range $1–1.6\ Nm^{-1}$. On the other hand, the bubble nucleation at gas-filled crevices on the container wall, or at the surface of solid inclusions in the liquid metal, occur with very small supersaturation.

As an example, let us consider the growth of CO bubbles in liquid steel for which the criterion is

$$P_{CO} > P_o + \frac{2\sigma}{r} \qquad (2.63)$$

where $P_{CO}$ is the pressure of CO in equilibrium with the concentrations of C and O in steel, $P_o$ the static pressure over the bubble and $r$ is the bubble radius. For the bubble to grow at the mouth of a cylindrical crevice, the radius of the crevice must be greater than the critical minimum $r_{min}$

$$r_{min} > \frac{2\sigma}{P_{CO} - P_o} \tag{2.64}$$

There is a maximum crevice radius above which the crevice cannot sustain the residual gas pocket, which is needed for bubble growth,

$$r_{max} < - \frac{2\sigma}{P_o} \cos \theta \tag{2.65}$$

where $\theta$ is the contact angle between the melt and the refractory surface; usually $\theta > 90°$.

### 2.6.2 GAS HOLDUP IN STEEL BATH

In most practical applications of gas injection into liquids, we are concerned mainly with rates of reactions in the swarm of bubbles. Mass-transfer phenomena in the swarm of bubbles have been studied extensively with aqueous solutions; accumulated knowledge is well documented in a review paper by Calderbank.[45] Also, several studies have been made of the bubble size and velocity of rise of single bubbles in liquid metals. Szekely presented a comprehensive review of these studies in his book on *Fluid Flow Phenomena in Metals Processing*.[46]

An attempt is made here to estimate the size range of dispersed gas bubbles in liquid steel for the conditions prevailing in BOF and Q–BOP steelmaking. The bubbles are assumed to be generated from a multitude of points across the bottom of the vessel, as though the entire vessel bottom were in fact one big porous plug.

The volume fraction of gas holdup ($\varepsilon$) is given by the following ratio:

$$\varepsilon = \frac{V_b}{V_b + V_s} = 1 - \frac{V_s}{HA} \tag{2.66}$$

where  $V_b$ = transitory bubble volume in the bath at any time,
  $V_s$ = steel volume in the vessel,
  $H$  = average bath depth,
  $A$  = average bath cross sectional area.
The velocity of bubble rise, $U_b$, in a swarm of bubbles is given by the ratio

$$U_b = \frac{U_s}{\varepsilon} = \frac{\dot{V}}{\varepsilon A} \tag{2.67}$$

where  $U_s$ = superficial gas velocity,
  $\dot{V}$ = gas flow rate.

Yoshida and Akita[47] have measured the volume fraction of gas holdup in non-foaming aqueous solutions with $U_s$ in the range 0.03 to 0.6 m s$^{-1}$. In these experiments, the gas was injected through a single nozzle on the bottom of the liquid column contained in a cylindrical vessel. From their experimental results for the air-water system, the following equation is obtained.

$$\log (1 - \varepsilon) = - 0.146 \log(1 + U_s) + 0.06 \tag{2.68}$$

In the studies of Calderbank and co-workers[45] with the air-water system, the gas was injected into the column of water through a sieve plate on the bottom of the column. The fractional holdup of gas was reported as shown in Fig. 2.29. The units in F factor are:

$$U_s, \text{ m s}^{-1} \text{ and } \rho_g \text{ (the gas density), kg m}^{-3}.$$

The experimental results of Crozier[48], not shown in Fig. 2.29, for F>0.3 are in good agreement with those reported by Calderbank. The dotted curve is calculated from equation (2.68), and plotted in Fig. 2.29 in terms of the F factor with $\rho_g$ (air at 25°C and 1 atm) = 1.185 kg m$^{-3}$. In the present assessment of gas holdup in the steel bath, the preference is given to the data reported by Calderbank. The dot-dash curve is taken to represent the average values. It appears that at $F \geq 1$, the gas holdup reaches an essentially constant value of $\varepsilon = 0.6$.

In the estimation of fractional gas holdup in the steel bath, during decarburisation or argon stirring, the gas flow rates taken are for the melt

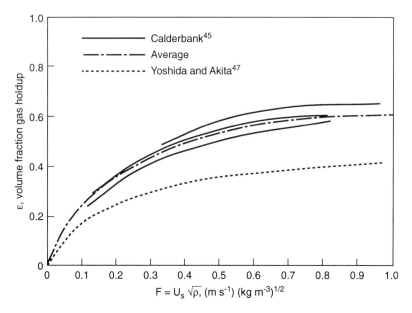

Fig. 2.29   Experimental data of Calderbank[45] on gas holdup in sieve plate columns for the air–water system are compared with those of Yoshida and Akita[47].

temperature and average gas bubble pressure. The values used are for 1600°C and 1.5 atm so that $\dot{V}_T(\text{m}^3\text{s}^{-1}) = 4.57 \times \dot{V}_o(\text{Nm}^3\text{s}^{-1})$; for these conditions the gas densities are 0.27 kg m$^{-3}$ for $N_2$ and CO and 0.39 kg m$^{-3}$ for Ar; an average value is used, $\rho_g = 0.33$ kg m$^{-3}$.

The average bath cross sectional area $A$ in BOF or Q–BOP vessel increases with an increasing gas flow rate. Keeping in mind the inner dimensions of the BOF or Q–BOP vessel for 200–240 tonne heats, the area values used are: $A = 22$ m$^2$ at $\dot{V}_T = 60$ m$^3$s$^{-1}$ decreasing to $A = 16$ m$^2$ at $\dot{V}_T = 1$ m$^3$s$^{-1}$. On this basis, the following equation is obtained for the superficial gas velocity $U_s$ as a function of $\dot{V}_T$.

$$U_s \text{ (m s}^{-1}) = \frac{\dot{V}_T}{14.55 + 0.089V_T} \tag{2.69}$$

The volume fraction gas holdup and bubble velocity of rise calculated using equation (2.69) and Fig. 2.29 are plotted in Fig. 2.30 as a function of the gas flow rate. The gas holdup in the slag layer is a more complex phenomenon, because of slag foaming and injection of large amounts of steel droplets into the slag layer during decarburisation with oxygen blowing.

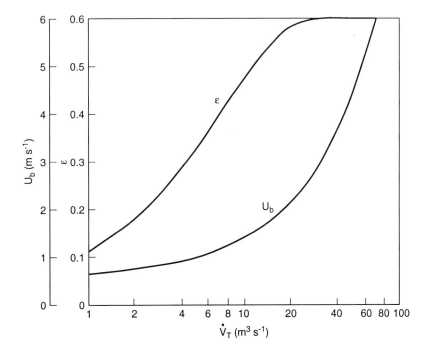

Fig. 2.30   Gas holdup and bubble velocity as a function of gas flow rate.

2.6.3 ESTIMATION OF BUBBLE SIZE FOR UNIFORMLY DISPERSED BUBBLES IN THE STEEL BATH

Many estimates were made in the past of gas bubble diameters in steel-making processes; the estimates varying over a wide range from 1 to 8 cm or more.*

Calderbank and co-workers[45] and Leibson et al.[49] have found that in aqueous solutions the bubble size becomes smaller with an increasing gas flow rate, ultimately reaching a minimum size of about 0.45 cm diameter at large gas flow rates. Relatively large gas bubbles in motion are subject to deformation and ultimately to fragmentation into smaller bubbles. The drag force exerted by the liquid on a moving bubble induces rotational and probably a turbulent motion of the gas within the bubble. This motion creates a dynamic pressure on the bubble surface; when this force exceeds the surface tension, bubble breakup occurs. Because of the large difference between the densities of the gas and liquid, the energy associated with the drag force is much greater than the kinetic energy of the gas bubble. There-fore, the gas velocity in the bubble will be similar to the bubble velocity. On the basis of this theoretical reasoning, Levich[50] derived the following equa-tion for the critical bubble size as a function of bubble velocity.

$$d_c = \left( \frac{3}{C_d\, \rho_g\, \rho_l^2} \right)^{1/3} \frac{2\sigma}{U_b^2} \qquad (2.70)$$

where $\sigma$ is the surface tension and $C_d$ the drag coefficient, which is close to unity. As the bubble size decreases with increasing bubble velocity, the gas circulation within the bubble diminishes, thus becoming less effective in bub-ble fragmentation. Another view to be considered is that, as the gas holdup increases with increasing gas flow rate, the liquid layer separating the bub-bles becomes thinner. Consequently, the dynamic gas pressure exerted on the bubble surface becomes nullified by similar forces exerted in the neighbour-ing bubbles, hence the cessation of bubble fragmentation at high values of $\varepsilon$.

The bubble fragmentation with increasing $\varepsilon$, hence increasing $U_b$, is calculated for bubbles in water with $\sigma = 0.072$ Nm$^{-1}$ and in liquid steel with (a) $\sigma = 1.7$ Nm$^{-1}$ in the presence of O < 4 ppm and 40 ppm S, and (b) $\sigma = 1.3$ Nm$^{-1}$ in the presence of 600 ppm O and 120 ppm S. The results are shown in Fig. 2.31. As noted earlier, the limiting bubble size in air-water system is $d = 0.45$ cm which intersects the fragmentation curve at $\varepsilon = 0.41$, depicting the cessation of bubble fragmentation.

In an attempt to estimate the minimum bubble size in the swarm of bubbles in liquid steel, the author proposes the following hypothesis: 'the

*The bubbles with diameters > 1 cm acquire spherical cap shape as they rise in liquids. The apparent bubble diameter is that of a sphere which has a volume equivalent to that of the spherical cap shape bubble.

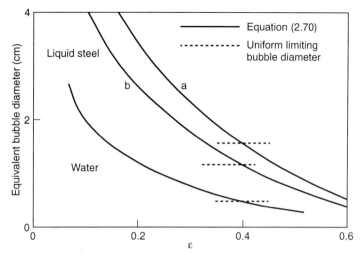

Fig. 2.31 Calculated bubble diameter for bubble fragmentation mechanism equation (2.70), compared with uniform limiting diameter at high ε values for water and liquid steel: (a) for O < 4 ppm and 40 ppm S (b) for 600 ppm O and 120 ppm S.

surface energy of bubbles per unit mass of liquid which the bubbles displace, are equal for all liquids.'[*] This conjectured statement leads to the following *similarity relation* for non-foaming inviscid liquids.

$$\frac{\sigma_1}{\rho_1 d_1} = \frac{\sigma_2}{\rho_2 d_2} = \dots \text{J kg}^{-1} \text{ liquid} \tag{2.71}$$

With $\sigma_1 = 0.072$ Jm$^{-2}$ and $d_1 = 0.45$ cm for gas-water system and $\sigma_2 = 1.3$ Jm$^{-2}$ and $\rho_2 = 6940$ kg m$^{-3}$, we obtain $d_2 = 1.16$ cm for the limiting bubble diameter in liquid steel containing $\approx 600$ ppm O and $\approx 120$ ppm S. For the melt containing O < 4 ppm and 40 ppm S, $d_2 = 1.52$ cm. These limiting values intersect the fragmentation curves for liquid steel at about ε = 0.40, similar to that for water.

A uniform constant bubble size and essentially constant ε = 0.6 at high gas flow-rates, means a constant number of bubbles per unit volume of the emulsion. Noting that for the close-packed (fcc) arrangement of bubbles, the packing fraction (gas holdup) is ε = 0.74, the thickness δ of the liquid film at the nearest approach of the bubbles is derived from the following equality of the bubble number.

$$\frac{0.6}{d^3} = \frac{0.74}{(d + \delta/2)^3} \tag{2.72}$$

Noting that for water the experimental data give $d = 0.45$ cm and that estimated for liquid steel $d = 1.16$ cm, equation (2.72) gives δ = 0.65 mm for

---

[*]Surface energy $E_b$ of dispersed bubbles of uniform size per unit mass of liquid which they displaced is given by

$$E_b = \frac{6\varepsilon}{1 - \varepsilon} \left(\frac{\sigma}{\rho d}\right) \quad \text{J kg}^{-1} \text{ liquid}$$

bubbles in aqueous solutions and $\delta = 1.68$ mm for bubbles in liquid steel at the limiting value of $\varepsilon = 0.6$ for non-foaming inviscid liquids.

2.6.4 RATE EQUATION FOR TRANSPORT CONTROLLED GAS BUBBLE REACTIONS IN LIQUID STEEL

For mass-transfer controlled reactions of gas bubbles in liquid steel, the rate equation has the following form.

$$\ln \frac{\%X - \%X_e}{\%X_o - \%X_e} = -S_o k_m t \tag{2.73}$$

where $X_o$ and $X_e$ are the initial and gas-metal equilibrium concentrations of the diffusing reactant in the melt, $k_m$ the mass-transfer coefficient and $S_o$ the bubble surface area which is in terms of $\varepsilon,$*

$$S_o = \frac{6\varepsilon}{d} \text{ , m}^2/\text{m}^3 \text{ gas-melt emulsion} \tag{2.74}$$

From many studies of gas bubble reactions in nonfoaming aqueous solutions, the following formulation has been derived for the liquid-phase mass-transfer coefficient for the regime of surface-renewal at the gas-liquid interface.[45]

$$k_m = 1.28 \left( \frac{DU_b}{d} \right)^{1/2} \tag{2.75}$$

where $D$ is the diffusivity of the reactant in the liquid-film boundary layer.

In the presence of surface active solutes such as oxygen and sulphur in liquid steel, the bubble surface will be less mobile hence the rate constant $k_m$ will be somewhat less than that given by equation (2.75) for the mobile surface, which is a necessary condition for surface-renewal. On the other hand, the surface active solutes decrease the bubble diameter, hence increase the bubble surface area for a given gas holdup. It appears that the product $S_o k_m$ may not be too sensitive to the presence of surface active solutes in the liquid.

The rate constant $S_o k_m$ for the transport controlled reaction is obtained from the combination of equations (2.67), (2.74) and (2.75), as given below for an average solute diffusivity $D = 5 \times 10^{-9}$ m$^2$s$^{-1}$.

$$S_o k_m = \frac{5.43 \times 10^{-4}(\varepsilon U_s)^{1/2}}{d^{3/2}} \tag{2.76}$$

*Bubble surface area with respect to unit mass of liquid is

$$S_m = \frac{6\varepsilon}{1-\varepsilon} \left( \frac{1}{\rho d} \right) \text{ , m}^2/\text{kg liquid}$$

The rate constant thus calculated is plotted in Fig. 2.32 against the gas flow rate $\dot{V}_T$ at 1600°C and for an average gas pressure of 1.5 atm in the steel bath for 220 ± 20 tonne heats.

### 2.6.5 AN EXAMPLE OF CALCULATED RATE OF NITROGEN REMOVAL

In Q–BOP steelmaking, the melt sometimes is purged (before tapping) with argon or an argon + oxygen mixture for 1 to 2 minutes at the rate of 2 $Nm^3min^{-1}t^{-1}$, usually for some hydrogen removal. In this example, we shall consider nitrogen removal controlled by two different mechanisms: (i) mass-transfer controlled and (ii) chemical reaction controlled.

Converting the gas flow rate to that in the melt at 1600°C and 1.5 atm pressure, for a 220 t heat, $\dot{V}_T = 34$ $m^3s^{-1}$ for which the rate constant $S_ok_m = 0.48$ $s^{-1}$ from Fig. 2.32. If the steel initially contains 30 ppm $N_o$ and the gas-metal equilibrium value $N_e$ is negligibly small, in 10 seconds of purging, the residual content will be 0.25 ppm N. It is all too clear from this example that, according to the present assessment of the rate phenomena, the transport of reactants to and from the gas bubbles in liquid steel are very fast at relatively high gas flow rates. It appears that the reaction rates would be controlled primarily either by an interfacial chemical reaction or by saturation of the gas bubbles with the reactant, or depletion thereof.

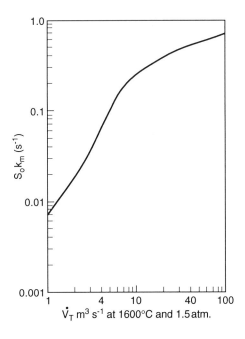

Fig. 2.32   Calculated rate constant for mass-transfer controlled reaction of gas bubbles in 220 ± 20 tonne steel bath in the BOF or Q–BOP vessel.

Now let us consider the chemical-reaction controlled rate of denitrogenation. From equation (2.25) in section 2.1.1a, the rate for 1600°C $\Phi_r$ = 0.233 g N cm$^{-2}$s$^{-1}$%N$^{-1}$. For liquid steel containing 600 ppm O and 120 ppm S, $(1 - \theta)$ = 0.055 from equation (2.21). For the gas flow rate $\dot{V}_T$ = 34 m$^3$s$^{-1}$, $\varepsilon$ = 0.6 and the bubble surface area is $S_m$ = 1.11 cm$^2$g$^{-1}$ liquid steel. Inserting these numbers in equation (2.22) for the limiting case of %N$_e \approx$ 0, gives

$$\frac{1}{\%N} - \frac{1}{\%N_o} = 100 \times 1.11 \times 0.233 \times 0.055t = 1.42t$$

In 60 seconds time of purging, 0.0030% N$_o$ will be reduced only to 0.0024% N. In actual fact, the argon bubbles traversing the melt will contain some N$_2$; for example, if the nitrogen partial pressure in gas bubbles were 0.001 atm N$_2$, the equilibrium content in the melt would be 0.0015% N$_e$. Therefore, with %N$_e$ > 0, the nitrogen removal will be much less than that calculated above for the chemical-reaction controlled rate for the hypothetical limiting case of %N$_e \approx$ 0.

The foregoing prediction from the rate equation is in complete accord with the practical experience that even at high rates of argon purging of the melt, as in Q–BOP, there is no perceptible nitrogen removal because of the presence of surface active solutes, oxygen and sulphur, in the steel bath.

## 2.7 MAXIMUM RATE OF DEGASSING OF LIQUID STEEL WITH ARGON PURGING

If the rates of diffusional processes and chemical reactions are sufficiently fast, the argon bubbles traversing the melt will be saturated with N$_2$ or H$_2$ to the corresponding equilibrium values for the concentrations present in the melt. It is for this limiting case that the following rate equation is applicable.

Denoting N and H by $X$, and using molar contents in the gas bubbles,

$$\dot{n}_X = \frac{d\,\text{ppm}X}{dt} \frac{10^{-6}}{M} \times 10^6 \text{ gmol } X_2/\text{min tonne}$$

$$\dot{n}_{Ar} = \frac{\dot{V}}{22.414 \times 10^{-3}} \text{ gmol Ar/min tonne}$$

where $M$ is the molecular mass of $X_2$ and $\dot{V}$ the argon blowing rate in Nm$^3$ min$^{-1}$t$^{-1}$. In the $X_2$-saturated argon bubbles for $\dot{n}_{Ar} \gg \dot{n}_{X2}$

$$p_{X2} = \frac{\dot{n}_{X2}}{\dot{n}_{Ar}} \bar{P} = \frac{d\,\text{ppm}X}{dt} \frac{22.414 \times 10^{-3}}{MV} \bar{P}$$

where $\bar{P}$ is the average bubble pressure in the melt.

Since the gas bubbles are in equilibrium with the melt

$$p_{X_2} = \frac{[\text{ppm } X]^2}{K^2} = \frac{d \text{ ppm}X}{dt} \frac{22.414 \times 10^{-3}}{M\dot{V}} \bar{P} \qquad (2.77)$$

where $K$ is the equilibrium constant. The integration of the above rate equation gives with $X_o$ being the initial concentration,

$$\frac{1}{\text{ppm}X} - \frac{1}{\text{ppm}X_o} = \frac{M\dot{V}}{22.414 \times 10^{-3}K^2\bar{P}} t \qquad (2.78)$$

For the average liquid steel temperature of 1650°C at turn down, the values of the equilibrium constants are as given below.

$$\left.\begin{array}{ll} \text{For } N_2\text{:} & K = 459 \\ \text{For } H_2\text{:} & K = 27.2 \end{array}\right\} \text{ with } p_{X_2} \text{ in atm}$$

For an average gas bubble pressure of $\bar{P} = 1.5$ atm in the melt, the following rate equations are obtained.

$$\frac{1}{\text{ppm N}} - \frac{1}{\text{ppm N}_o} = 0.00395 \, \dot{V}t \qquad (2.79)$$

$$\frac{1}{\text{ppm H}} - \frac{1}{\text{ppm H}_o} = 0.0804 \, \dot{V}t \qquad (2.80)$$

As shown in the previous section, the nitrogen removal by argon purging even at high rates is not possible because of the low rate of chemical reaction at the melt-bubble surface in steels containing relatively large concentrations of oxygen and sulphur at tap. If the reaction kinetics are favourable, some hydrogen removal may be achieved at very high argon flow rates as in Q–BOP. For example, purging with argon at the rate of 2 Nm³ min⁻¹ t⁻¹ steel for 2 minutes, at the maximum rate of degassing, the initial 7 ppm H would be reduced to 2 ppm.

From the observations made at the Gary Q-BOP Shop of U.S. Steel, it was noted that there was very little hydrogen removal with argon purging as given in the above example. However, with an argon + oxygen mixture, there was about 3 to 4 ppm H removal in 2 minutes purging at the rate of 2 Nm³min⁻¹t⁻¹, which is about 75% of the maximum possible rate. These observations indicate that with Ar + $O_2$ purging, the rate of the reaction

$$2[\text{H}] + \frac{1}{2} O_2(g) \rightarrow H_2O(g)$$

is considerably faster than the rates of either of the following reactions with argon alone.

$$2[\text{H}] \rightarrow H_2(g)$$

$$2[\text{H}] + [\text{O}] \rightarrow H_2O(g)$$

## 2.8 COMPUTER-BASED PACKAGES

There are numerous computer software packages, as for example PHOENICS, FLUENT, CFDS-FLOWS3D, GENMIX, TEACH(2D, 3D), 2E/FIX, SOLA/VOF, METADEX(R) and FID, which are used in the steel industry in the process and design engineering of plant facilities and in the control of steelmaking processes. References may also be made to the publications by Szekely *et al.*[51–53] on the mathematical and physical modelling of metallurgical processes involving fluid flow, heat transfer, mass transfer and gas–slag–metal reactions.

There are of course many other publications on this subject which cannot all be cited here as the subject matter is well outside the scope of this book.

## REFERENCES

1. J. SZEKELY and N.J. THEMELIS, *Rate phenomena in process metallurgy*, Wiley, New York, 1971.
2. G.H. GEIGER and D.R. POIRIER, *Transport phenomena in metallurgy*, Addison-Wesley, Reading, Massachusetts, 1973.
3. S. GLASSTONE, K.J. LAIDLER and H. EYRING, *The theory of rate processes*, McGraw-Hill, New York, 1941.
4. C.N. HINSHELWOOD, *Kinetics of chemical change*, Oxford University Press, London, 1942.
5. T. FUWA, S. BAN-YA and T. SHINOHARA, *Tetsu-to-Hagané*, 1967, **53**, S328.
6. M. INOUYE and T. CHOH, *Trans. Iron Steel Inst. Japan*, 1968, **8**, 134.
7. K. MORI and K. SUZUKI, *Trans. Iron Steel Inst. Japan*. 1970, **10**, 232.
8. R.J. FRUEHAN and L.J. MARTONIK, *Metall. Trans. B*, 1980, **11B**, 615.
9. M. BYRNE and G.R. BELTON, *Metall. Trans. B*, 1983, **14B**, 441.
10. S. BAN-YA, T. SHINOHARA, H. TOZAKI and T. FUWA, *Tetus-to-Hagané*, 1974, **60**, 1443.
11. D.R. SAIN and G.R. BELTON, *Metall. Trans. B*, 1976, **7B**, 235; 1978, **9B**, 403.
12. A.C. SMIGELSKAS and E.O. KIRKENDALL, *Trans. AIME*, 1947, **171**, 130.
13. J.P. STARK, *Acta Metall.*, 1966, **14**, 228.
14. L.S. DARKEN, *Trans. AIME*, 1948, **174**, 184.
15. G.S. HARTLEY and J. CRANK, *Trans. Faraday Soc.* 1949, **76**, 1169.
16. J.S. SHEASBY, W.E. Boggs and E.T. TURKDOGAN, *Metal Science*, 1984, **18**, 127.
17. C. WAGNER, *Z. Phys. chem.*, Abt B, 1933, **21**, 25; in *Atom movements*, p. 153. ASM, 1950.
18. L. HIMMEL, R.F. MEHL and C.E. BIRCHENALL, *Trans. AIME*, 1953, **197**, 827.
19. P. HEMBREE and J.B. WAGNER, *Trans. AIME*, 1969, **245**, 1547.
20. E.T. TURKDOGAN, R.G. OLSSON, H.A. WRIEDT and L.S. DARKEN, *SME Trans.*, 1973, **254**, 10.
21. E.H. BAKER, *J. Chem. Soc.*, 1962, 464.
22. A.W.D. HILLS, *Inst. Min. Metall. Trans.*, 1967, **76**, 241.

23. E.W. Thiele, *Ind. Eng. Chem.*, 1939, **31**, 916.
24. A. Wheeler, *Catalysis*, 1955, **2**, 105.
25. P.B. Weisz and C.D. Prater, *Advan Catalysis*, 1954, **6**, 143.
26. P.L. Walker, F. Rusinko and L.G. Austin, *Advan. Catalysis*, 1959, **21**, 33.
27. E.T. Turkdogan, V. Koump, J.V. Vinters and T.F. Perzak, *Carbon*, 1968, **6**, 467.
28. E.T. Turkdogan and J.V. Vinters, *Carbon*, 1969, **7**, 101; 1970, **8**, 39.
29. E.T. Turkdogan, R.G. Olsson and J.V. Vinters, *Carbon*, 1970, **8**, 545.
30. E.A. Gulbransen, K.F. Andrew and F.A. Brassort, *Carbon*, 1965, **2**, 421.
31. J.D. Blackwood, *Aust. J. Appl. Sci.*, 1962, **13**, 199.
32. S. Ergun, *J. Phys. Chem.*, 1956, **60**, 480.
33. K. Hedden and A. Löwe, *Carbon*, 1967, **5**, 339.
34. H.-J. Grabke, *Ber. Bunsenges. physik Chem.*, 1966, **70**, 664.
35. D.A. Aderibigbe and J. Szekely, *Ironmaking & Steelmaking*, 1981, **8**, 11.
36. R.H. Tien and E.T. Turkdogan, *Carbon*, 1970, **8**, 607.
36a. H. Jung, R.S. Armstead, N. Al. Ibrahim and B. Bowman, *Ironmaking & Steelmaking*, 1990, **17**, 118.
37. W.M. McKewan, *Trans. TMS-AIME*, 1960, **218**, 2.
38. J.M. Quets, M.E. Wadsworth and J.R. Lewis, *Trans. TMS-AIME*, 1960, **218**, 545: 1961, **221**, 1186.
39. N.J. Themelis and W.H. Gauvin, *A.I.Ch.E.J.*, 1962, **8**, 437; *Trans. TMS-AIME*, 1963, **227**, 290.
40. N.A.Warner, *Trans. TMS-AIME*, 1964, **230**, 163.
41. R.H. Spitzer, F.S. Manning and W.O. Philbrook, *Trans. TMS-AIME*, 1966, **236**, 726.
42. E.T. Turkdogan and J.V. Vinters, *Metall. Trans.*, 1971, **2**, 3175; 1972, **3**, 1561.
43. E.T. Turkdogan, R.G. Olsson and J.V. Vinters, *Metall. Trans.*, 1971, **2**, 3189.
44. E.T. Turkdogan, P. Grieveson and L.S. Darken, *J. Phys. Chem.*, 1963, **67**, 1647; *J. Metals*, 1962, **14**, 521.
45. P.H. Calderbank, *Chem. Eng.* (London), 1967, **212**, CE 209.
46. J. Szekely, *Fluid flow phenomena in metals processing*, Academic Press, New York, 1979.
47. F. Yoshida and K. Akita, *A.I.Ch.E.J.*, 1965, **11**, 9.
48. R.D. Crozier, Ph.D. Thesis, University of Michigan, Ann Arbor, 1956.
49. I. Leibson, E.G. Holcomb, A.G. Cacuso and J.J. Jacmic, *A.I.Ch.E.J.*, 1956, **2**, 296.
50. V.G. Levich, *Physicochemical hydrodynamics*, Prentice-Hall, Englewood, Cliff., New Jersey, 1962.
51. J. Szekely, J.W. Evans and J.K. Brimacombe. The mathematical and physical modelling of primary metals processing operations. Wiley, New York, 1988.
52. J. Szekely, G. Carlsson and L. Helle, Ladle metallurgy. Springer, Berlin, 1989.
53. J. Szekely and O.J. Ilegbusi, The physical and mathematical modelling of tundish operations. Springer, Berlin, 1989.

CHAPTER 3

# Thermochemical and Transport Properties of Gases

Gases play a key role in all phases of pyrometallurgical processes. In assessing the technical aspects of ironmaking and steelmaking and the new developments on these processes, some information is always needed on the physical and chemical properties of gases at elevated temperatures. Therefore, it would be appropriate to give in this chapter a selected set of data and other relevant information on the thermochemical and transport properties of gases.

## 3.1 THERMOCHEMICAL PROPERTIES

### 3.1.1 MOLAR HEAT CAPACITY

The molar heat capacities of a few simple gases are plotted in Fig. 3.1. With the exception of $SO_2$ and $CO_2$, for other gases $C_P$ increases almost linearly with an increasing temperature. For monatomic gases $C_P$ is essentially independent of temperature.

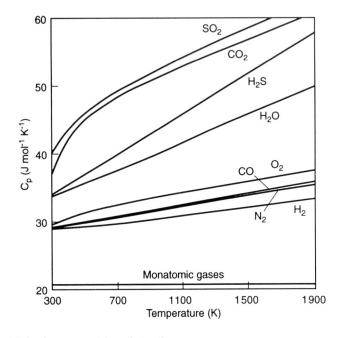

Fig. 3.1   Molar heat capacities of simple gases.

There is an interesting correlation for the heat capacity per atom, i.e. $\bar{C}_P$ = molar $C_P$/number of atoms per mol, as shown below.

| number of atoms per mol | $\bar{C}_P$ J g-atom$^{-1}$K$^{-1}$at 298K |
|:---:|:---:|
| 1 | 20.8 ± 0.05 |
| 2 | 14.6 ± 0.2 |
| 3–6 | 12.2 ± 1.0 |

The average of these values $\bar{C}_P = 16 \pm 4$ is approximately similar to that for complex crystalline minerals and aluminosilicate glasses discussed in section 1.6 of Chapter 1.

### 3.1.2 HEAT CONTENT

The molar heat contents of simple gases are plotted in Fig. 3.2 over the range 300–1900K.

### 3.1.3 EQUILIBRIUM STATES IN GAS MIXTURES

When assessing the state of reactions involving gas mixtures, the partial pressures of the gaseous species should be evaluated from the reaction equilibrium data for all the pertinent gas reactions. The method of calculation is demonstrated below by two examples.

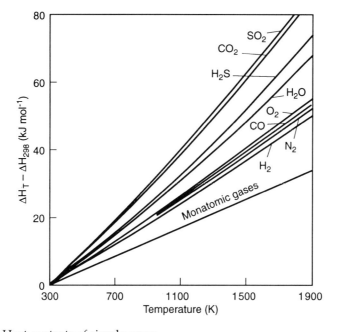

Fig. 3.2   Heat contents of simple gases.

A. Gas mixture $CO–CO_2–SO_2$ :

At elevated temperatures the following are the predominant gaseous species in the equilibrated gas mixture: $CO_2$, $CO$, $COS$, $S_2$ and $SO_2$: other species such as $O_2$, $S$, $SO$, $SO_3$, $CS$ and $CS_2$ are small enough to be neglected in the calculations. With this simplification, the material balance for carbon ($\Sigma C$), sulphur ($\Sigma S$) and oxygen ($\Sigma O$) are represented by the following ratios for the ingoing gas mixture at $(p_{CO})_i + (p_{CO_2})_i + (p_{SO2})_i = 1$ atmosphere and ambient temperature.

$$\frac{\Sigma C}{\Sigma S} = \frac{1-(p_{SO_2})_i}{(p_{SO_2})_i} = \left(\frac{p_{CO_2} + p_{CO} + p_{COS}}{2p_{S_2} + p_{SO_2} + p_{COS}}\right)_e \tag{3.1}$$

$$\frac{\Sigma C}{\Sigma O} = \frac{1-(p_{SO_2})_i}{2-(p_{CO})_i} = \left(\frac{p_{CO_2} + p_{CO} + p_{COS}}{2p_{SO_2} + p_{COS} + 2p_{CO_2} + p_{CO}}\right)_e \tag{3.2}$$

where e indicates partial pressures in the equilibrated gas mixture.

For the reaction $2CO + SO_2 = \frac{1}{2}S_2 + 2CO_2$, $\tag{3.3}$

$$K_S = \left(\frac{p_{CO_2}}{p_{CO}}\right)^2 \frac{(p_{S_2})^{1/2}}{p_{SO_2}} \tag{3.3a}$$

For the reaction $3CO + SO_2 = COS + 2CO_2$, $\tag{3.4}$

$$K_C = \frac{(p_{CO_2})^2}{(p_{CO})^3} \frac{p_{COS}}{p_{SO_2}} \tag{3.4a}$$

For the sum of all the predominant species,

$$(p_{CO_2} + p_{CO} + p_{COS} + p_{S_2} + p_{SO_2})_e = 1 \tag{3.5}$$

By simultaneous solution of these five equations, for known values of the equilibrium constants $K_S$ and $K_C$ and known ingoing gas composition, the equilibrium partial pressures of the gaseous species are calculated through a computer programme. Examples of the calculated data are given in Fig. 3.3 showing the equilibrium ratio $p_{CO_2}/p_{CO}$ at the indicated temperatures as a function of percent $(SO_2)_i$ in the ingoing gas.

B. Gas mixture $CO–CO_2–H_2O–CH_4$:

This example is in relation to reheating a reformed gas in the direct iron ore reduction processes. If there is complete gas equilibrium at the reheating temperature and pressure, the activity of carbon in the equilibrated gas mixture must satisfy the following three basic reaction equilibria:

$$2CO = C + CO_2 \tag{3.6}$$
$$K_1 = a_C p_{CO_2}/p_{CO}^2 \tag{3.6a}$$

$$H_2 + CO = C + H_2O \tag{3.7}$$
$$K_2 = a_C p_{H_2O}/p_{H_2}p_{CO} \tag{3.7a}$$

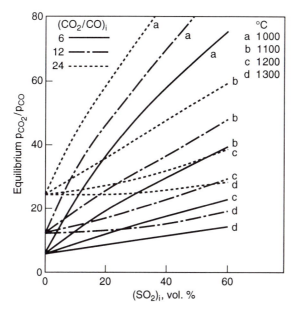

Fig. 3.3 Equilibrium ratios $p_{CO_2}/p_{CO}$ at indicated temperatures as a function of pct $(SO_2)_i$ for indicated ratios $(CO_2/CO)_i$ in ingoing gas mixtures $CO$–$CO_2$–$SO_2$ at atmospheric pressure.

$$CH_4 = C + 2H_2 \tag{3.8}$$
$$K_3 = a_C p_{H2}^2 / p_{CH4} \tag{3.8a}$$

For the predominant gaseous species the following equalities are derived from the mass balance for $(\Sigma C)$, $(\Sigma O)$ and $(\Sigma H)$.

$$\frac{\Sigma C}{\Sigma O} = \left(\frac{p_{CO_2} + p_{CO} + p_{CH4}}{2p_{CO_2} + p_{CO} + p_{H_2O}}\right)_i = \left(\frac{p_{CO_2} + p_{CO} + p_{CH4}}{2p_{CO_2} + p_{CO} + p_{H_2O}}\right)_e \tag{3.9}$$

$$\frac{\Sigma C}{\Sigma H} = \left(\frac{p_{CO_2} + p_{CO} + p_{CH4}}{2p_{H_2O} + 2p_{H_2} + 4p_{CH4}}\right)_i = \left(\frac{p_{CO_2} + p_{CO} + p_{CH4}}{2p_{H_2O} + 2p_{H_2} + 4p_{CH4}}\right)_e \tag{3.10}$$

where $i$ and $e$ indicate partial pressures in the ingoing and equilibrated gas mixtures, respectively. The equation below gives the total pressure $P$ of the equilibrated gas mixture.

$$(p_{CO} + p_{CO_2} + p_{H_2} + p_{H_2O} + p_{CH_4})_e = P \tag{3.11}$$

The partial pressures of gaseous species and the activity of carbon in the equilibrated gas mixture are computed by simultaneous solution of equations (3.6) to (3.11) with known values of the equilibrium constants $K_1$, $K_2$ and $K_3$.

Examples are given in Fig. 3.4 showing the calculated carbon activities for 4 atm total pressure and for an initial gas mixture containing 73% $H_2$, 18% $CO$, 8% $CO_2$, 1% $CH_4$ to which 0.2% to 6.0% $H_2O$ has been added. The carbon deposition is imminent when its activity in the equilibrated gas

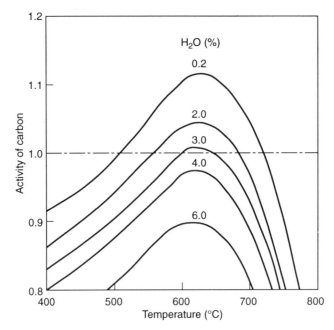

Fig. 3.4   Calculated activity of carbon for complete gas equilibrium at 4 atm and indicated temperatures for reformed natural gas containing 73% $H_2$, 18% CO, 8% $CO_2$ and 1% $CH_4$ to which an indicated amount of $H_2O$ is added.

exceeds unity, with respect to graphite. With the addition of 0.2% $H_2O$ to this gas mixture, there should be no carbon deposition in the equilibrated gas mixture at temperatures below 512°C and above 720°C. In an earlier study of the reduction of iron oxides in a similar gas mixture, carbon deposition was observed at all temperatures below 1000°C (see Ref. 1), indicating lack of complete gas equilibrium even at 1000°C.

### 3.2 TRANSPORT PROPERTIES

The molecular transfer of mass, momentum and energy are interrelated transport processes of diffusion under a concentration gradient, viscous flow in a velocity gradient and heat conduction in a thermal gradient. The earlier derivation of equations for the transport properties, as given below, were based on an oversimplified kinetic theory of gases.

$$D = \tfrac{1}{3}\bar{v}l \qquad \text{(diffusivity)} \qquad (3.12)$$
$$\eta = \tfrac{1}{3}nm\bar{v}l \qquad \text{(viscosity)} \qquad (3.13)$$
$$\kappa = \tfrac{1}{3}nc_v\bar{v}l \qquad \text{(thermal conductivity)} \qquad (3.14)$$

where   $m$ = molecular mass
$\quad\quad n$ = number of molecules per cm³
$\quad\quad \bar{v}$ = mean thermal molecular velocity
$\quad\quad l$  = mean free path
$\quad\quad c_v$ = specific heat capacity per g at constant volume.

The derivation of the transport properties from the rigorous kinetic theory of gases, described in depth by Chapman and Cowling[2] and Hirschfelder *et al.*[3] is based on the evaluation of the intermolecular energy of attraction, the collision diameter and the collision integral involving the dynamics of a molecular encounter, hence the intermolecular force law.

For a brief description of the theoretical equations for the calculation of the transport properties of gases, reference may be made to the author's previous publication.[4] For the present purpose, it is sufficient to give selected numerical data on the transport properties of a few gases, relevant to ironmaking and steelmaking processes.

**Table 3.1**   Diffusivity $D$ at 1 atm pressure: $cm^2s^{-1} \equiv 10^{-4}m^2s^{-1}$

| | Temperature, K | | | |
|---|---|---|---|---|
| Gas | 300 | 1000 | 1500 | 2000 |
| $H_2$ | 1.46 | 10.82 | 21.07 | 33.90 |
| Ar | 0.18 | 1.46 | 2.88 | 4.56 |
| $N_2$ | 0.20 | 1.59 | 3.05 | 4.98 |
| $O_2$ | 0.22 | 1.67 | 3.27 | 5.30 |
| CO | 0.20 | 1.63 | 3.20 | 5.19 |
| $CO_2$ | 0.11 | 0.93 | 1.87 | 3.00 |
| $H_2O$ | 0.28 | 2.83 | 5.79 | 9.57 |

**Table 3.2**   Viscosity $\eta$: (poise) $\equiv g\ cm^{-1}s^{-1} \equiv 0.1N\ s\ m^{-2} \equiv 0.1J\ m^{-3}s$

| | Temperature, K | | | |
|---|---|---|---|---|
| Gas | 300 | 1000 | 1500 | 2000 |
| $H_2$ | 0.89 | 1.95 | 2.53 | 3.06 |
| Ar | 2.50 | 5.34 | 6.98 | 8.27 |
| $N_2$ | 1.77 | 4.07 | 5.11 | 6.30 |
| $O_2$ | 2.07 | 4.82 | 6.20 | 7.64 |
| CO | 1.79 | 4.15 | 5.34 | 6.53 |
| $CO_2$ | 1.49 | 3.85 | 5.01 | 6.05 |
| $H_2O$ | 1.57 | 4.77 | 6.44 | 7.84 |

**Table 3.3**   Thermal conductivity $\kappa \times 10^4$, $J\ cm^{-1}s^{-1}k^{-1} \equiv \times 10^2$, $kg\ m\ s^{-3}K^{-1} \equiv \times 10^2$, $Wm^{-1}K^{-1}$

| | Temperature, K | | | |
|---|---|---|---|---|
| Gas | 300 | 1000 | 1500 | 2000 |
| $H_2$ | 17.45 | 39.96 | 53.87 | 67.64 |
| Ar | 1.97 | 4.18 | 5.44 | 6.44 |
| $N_2$ | 2.50 | 6.18 | 8.15 | 10.53 |
| $O_2$ | 2.57 | 6.68 | 9.02 | 11.62 |
| CO | 2.53 | 6.35 | 8.57 | 10.96 |
| $CO_2$ | 1.62 | 5.49 | 7.71 | 9.95 |
| $H_2O$ | 3.84 | 13.55 | 20.20 | 26.93 |

### 3.2.1 INTERRELATIONS BETWEEN TRANSPORT PROPERTIES

*3.2.1a Viscosity/thermal conductivity*

For monatomic gases,

$$\kappa = \frac{15R}{4M}\,\eta \tag{3.15}$$

where  $R = 8.314$ J mol$^{-1}$K$^{-1}$ (molar gas constant)
   $M$ = molecular mass, kg mol$^{-1}$
   $\eta$ = viscosity, N s m$^{-2}$ $\equiv$ J m$^{-3}$s
For polyatomic gases,

$$\kappa = \frac{15R}{4M}\,\eta\left\{\frac{4}{15}\frac{C_V}{R} + \frac{3}{5}\right\} \tag{3.16}$$

where $C_V$ is the molar heat capacity at constant volume and $\left\{\frac{4}{15}\frac{C_V}{R} + \frac{3}{5}\right\}$ is the 'Eucken[5] correction factor' for thermal conductivity of polyatomic gases. For monatomic gases $C_V = 3R/2$; with this substitution, equation (3.16) is reduced to equation (3.15). In terms of the molar heat capacity at constant pressure, $C_V = C_P - R$, equation (3.16) is transformed to the following form.

$$\kappa = \left( C_P + \frac{5R}{4} \right)\frac{\eta}{M} \tag{3.17}$$

*3.2.1b Thermal diffusivity/thermal conductivity*

The thermal diffusivity, $D^T$, is analogous to mass diffusivity and is given by the ratio

$$D^T = \frac{\kappa}{\rho C_P} \tag{3.18}$$

where $\rho$ is the molar density of the gas.

*3.2.1c Temperature and pressure effects*

The diffusivity, viscosity and thermal conductivity increase with an increasing temperature, thus

$$D \propto T^{3/2} \tag{3.19}$$

$$\eta \propto T^{1/2} \tag{3.20}$$

$$\kappa \propto T^{1/2} \tag{3.21}$$

The viscosity and thermal conductivity are independent of pressure; however, the diffusivity is inversely proportional to pressure.

*3.2.1d Molecular mass effect*

$$D \propto M^{-1/2} \tag{3.22}$$

$$\eta \propto M^{1/2} \tag{3.23}$$

$$\kappa \propto M^{-1} \tag{3.24}$$

## 3.3 PORE DIFFUSION

When the pores are small enough such that the mean free path $l$ of the molecules is comparable to the dimensions of the pore, diffusion occurs via the collision of molecules with the pore walls. Knudsen[6] showed that the flux involving collision of molecules with reflection from the surface of the capillary wall is represented by the equation

$$J = - \frac{2}{3} \bar{v} r \frac{dC}{dx} \tag{3.25}$$

where $\bar{v}$ is the mean thermal molecular velocity $= (8kT/\pi m)^{1/2}$, $r$ the radius of capillary tube and $dC/dx$ the concentration gradient along the capillary. For $r$ in cm and the Knudsen diffusivity $D_K$ in cm²s⁻¹

$$D_K = 9.7 \times 10^3 r \left( \frac{T}{M} \right)^{1/2} \tag{3.26}$$

The Knudsen diffusivity being independent of pressure but molecular diffusivity inversely proportional to pressure, it follows that in any porous medium, the Knudsen diffusion predominates at low pressures where the mean free path is larger than the dimensions of the pore.

For the mixed region of molecular and Knudsen diffusion in capillaries, Bosanquet[7] derived the following expression based on a random walk model in which the successive movements of molecules are terminated by collision with the capillary wall or with other molecules:

$$\frac{1}{D} = \frac{1}{D_{Ki}} - \frac{1}{D_{ii}} \tag{3.27}$$

where $D_{ii}$ is the molecular self-diffusivity. For a porous medium of uniform pore structure with pores of equal size, the Bosanquet interpretation formula gives for the effective diffusivity of component $i$

$$D_{ei} = \frac{\varepsilon}{\tau} \frac{D_{12} D_{Ki}}{D_{12} + D_{Ki}} \tag{3.28}$$

where $\varepsilon$ is the volume fraction of connected pores, $\tau$ the tortuosity factor, $D_{12}$ the molecular diffusivity for a binary mixture 1–2 and $D_{Ki}$ the Knudsen diffusivity of component $i$ for a given uniform pore radius $r$.

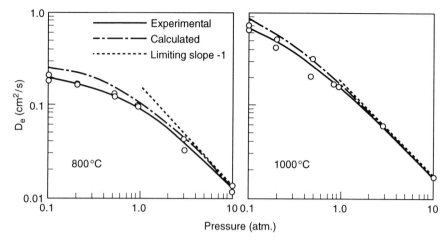

Fig. 3.5   Pressure dependence of effective diffusivity He–$CO_2$ at 20°C in iron reduced from haematite ore in hydrogen at indicated temperature.

The experimental data are shown in Fig. 3.5 for effective diffusivity He–$CO_2$ at 20°C in porous iron reduced from haematite ore in hydrogen at 800 and 1000°C.[8] As discussed in section 2.4.4, the pore structure of iron becomes finer at a lower reduction temperature. It is for this reason that the Knudsen diffusion effect at lower pressures become more pronounced for porous iron reduced at 800°C as compared to that reduced at 1000°C. More detailed information is given in the author's previous publications on the subject of gas diffusion in porous media.[4,8]

## REFERENCES

1. E.T. TURKDOGAN and J.V. VINTERS, *Metall. Trans.*, 1974, **5**, 11.
2. S. CHAPMAN and T.G. COWLING, *The mathematical theory of non-uniform gases*, Cambridge Univ. Press, London, (3rd edn. 1970).
3. J.O. HIRSCHFELDER, C.F. CURTISS and R.B. BIRD, *Molecular theory of gases and liquids*, Wiley, New York, 1954.
4. E.T. TURKDOGAN, *Physical chemistry of high temperature technology*, Academic Press, New York, 1980.
5. A. EUCKEN, *Physik.Z.*, 1913, **14**, 324.
6. M. KNUDSEN, *Ann. Phys.* (Leipzig), 1909, **28**, 75.
7. C.H. BOSANQUET, Br. TA Rep. BR-507, September 1944.
8. E.T. TURKDOGAN, R.G. OLSSON and J.V. VINTERS, *Metall. Trans.*, 1971, **2**, 3189.

# CHAPTER 4

# Physicochemical Properties of Steel

## 4.1 SELECTED THERMODYNAMIC ACTIVITIES

### 4.1.1 ACTIVITY COEFFICIENTS IN FE-X BINARY MELTS

In binary metallic solutions with complete liquid miscibility, there are positive or negative departures from Raoult's law and Henry's law as shown schematically in Fig. 4.1.

For low solute contents, as in low-alloy steels, the activity is defined with respect to Henry's law and mass percent of the solute

$$a_i = f_i \times [\%i] \qquad (4.1)$$

such that $f_i \rightarrow 1.0$ as $[\%i] \rightarrow 0$.

Up to several percent of the solute content, $\log f_i$ increases or decreases linearly with an increasing solute concentration.

$$\log f_i = e_i[\%i] \qquad (4.2)$$

The proportionality factor $e_i$ is known as the interaction coefficient.

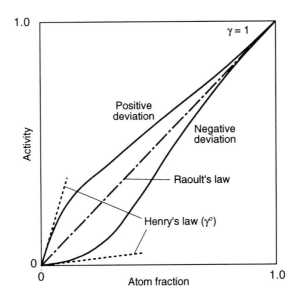

Fig. 4.1   Shematic representation of nonideal behaviour in binary metallic solutions.

**Table 4.1**  Values of $e_i^i$ for dilute solutions in liquid iron at 1600°C

| Element $i$ | $e_i^i$ | Element $i$ | $e_i^i$ |
|---|---|---|---|
| Al | 0.045 | O | − 0.10 |
| C | 0.18 | P | 0.062 |
| Cu | 0.023 | S | − 0.028 |
| Cr | 0.0 | Si | 0.11 |
| Mn | 0.0 | Ti | 0.013 |
| Ni | 0.0 | V | 0.015 |

For dilute solutions in liquid iron, the binary interaction coefficients $e_i^i$ are listed in Table 4.1, taken from the data compiled by Sigworth and Elliott.[1] Using the experimental data of Smith,[1a] the carbon activity is plotted in Fig. 4.2 with respect to two different standard states. When the standard state is chosen so that in infinitely dilute solutions the activity coefficient $f_C$ approaches unity, the temperature effect on the activity vs composition relation becomes almost negligible, at least within the range 800°–1200°C.

The activity of carbon in liquid iron was measured in many independent studies. The data compiled and re-assessed by Elliott *et al.*[2] are given in Figs. 4.3 and 4.4 as activity coefficients $\gamma_C$ and $f_C$ for two different standard states.

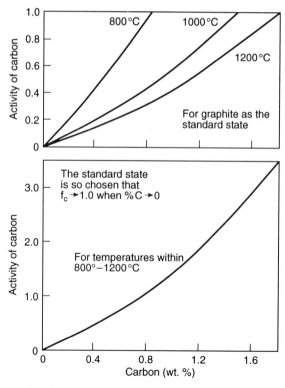

Fig. 4.2  Activity of carbon in austenite for two different standard states. From Ref. 1a.

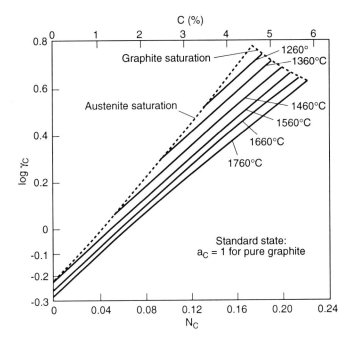

Fig. 4.3   Activity coefficient ($\gamma_C$) of carbon in liquid iron.

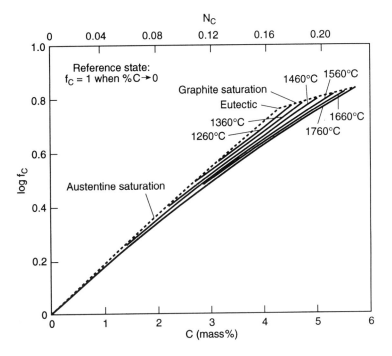

Fig. 4.4   Activity coefficient ($f_C$) of carbon in liquid iron.

4.1.2 ACTIVITY COEFFICIENTS IN MULTICOMPONENT MELTS

In multicomponent melts, as in alloy steels, the activity coefficient of solute i is affected by the alloying elements for which the formulation is

$$\log f_i = e_i^i\,[\%i] + \Sigma e_i^j \times [\%j] \tag{4.3}$$

where $e_i^j$ is the interaction coefficient of i as affected by the alloying element j.

$$e_i^j = \frac{d\log f_i}{d[\%j]} \tag{4.4}$$

Selected interaction coefficients in dilute solutions of ternary iron base alloys for C, H, N, O and S at 1600°C are given in Table 4.2.

**Table 4.2**   Selected interaction coefficients in dilute solutions of ternary iron base alloys for C, H, N, O and S at 1600°C, from Ref. 1

| Element j | $e_C^j$ | $e_H^j$ | $e_N^j$ | $e_O^j$ | $e_S^j$ |
|---|---|---|---|---|---|
| Al | 0.043 | 0.013 | −0.028 | −3.9 | 0.035 |
| C | 0.14 | 0.06 | 0.13 | −0.13 | 0.11 |
| Cr | −0.024 | −0.002 | −0.047 | −0.04 | −0.011 |
| Mn | −0.012 | −0.001 | −0.02 | −0.021 | −0.026 |
| N | 0.13 | 0 | 0 | 0.057 | 0.007 |
| O | −0.34 | −0.19 | 0.05 | −0.20 | −0.27 |
| P | 0.051 | 0.011 | 0.045 | 0.07 | 0.29 |
| S | 0.046 | 0.008 | 0.007 | −0.133 | −0.028 |
| Si | 0.08 | 0.027 | 0.047 | −0.131 | 0.063 |

For the mass concentrations of carbon and silicon above 1%, as in the blast furnace and foundry irons, the following values of $f_S^i$ should be used.

| Mass % C or Si : | 2.0 | 2.5 | 3.0 | 3.5 | 4.0 | 4.5 | 5.0 |
|---|---|---|---|---|---|---|---|
| $f_S^C$ : | 1.79 | 2.14 | 2.53 | 3.05 | 3.74 | 4.56 | 5.75 |
| $f_S^{Si}$ : | 1.37 | 1.50 | 1.64 | 1.78 | 1.95 | 2.10 | 2.32 |

4.1.3 FREE ENERGIES OF SOLUTION IN LIQUID IRON

For the solution of element $X_i$ in liquid iron at 1 mass % $X_i$,

$$X_i(\text{pure}) = [X_i]\ (1\text{ mass }\%)$$

the free energy of solution is

$$\Delta G_s = RT\ \ln\frac{0.5585}{M_i}\ \gamma_i^o$$

where $M_i$ is the atomic mass (g-atom) and $\gamma_i^o$ the activity coefficient (with respect to pure element) at infinite dilution ($\%X_i \equiv 0$).

For $\Delta G_s$ in J mol$^{-1}$ and substituting log for ln,

$$\Delta G_s = 19.146\, T \log \frac{0.5585}{M_i}\, \gamma_i^o \qquad (4.5)$$

The free energies of solution of various elements in liquid iron are listed in Table 4.3.

**Table 4.3**   Free energies of solution in liquid iron for 1 mass %: (*g*) gas, (*l*) liquid, (*s*) solid, from Ref. 1

| Element i | $\gamma_i^o$ | $\Delta G_s$, J mol$^{-1}$ |
|---|---|---|
| Al(*l*) | 0.029 | −63,178 − 27.91T |
| C (*gr*) | 0.57 | 22,594 − 42.26T |
| Co(*l*) | 1.07 | 1,004 − 38.74T |
| Cr(*s*) | 1.14 | 19,246 − 46.86T |
| Cu(*l*) | 8.60 | 33,472 − 39.37T |
| ½H$_2$(*g*) | – | 36,377 + 30.19T |
| Mg(*g*) | – | −78,690 + 70.80T |
| Mn (*l*) | 1.30 | 4,084 − 38.16T |
| ½N$_2$(*g*) | – | 3,599 + 23.74T |
| Ni(*l*) | 0.66 | −20,920 − 31.05T |
| ½O$_2$(*g*) | – | −115,750 −  4.63T |
| ½P$_2$(*g*) | – | −122,173 − 19.25T |
| ½S$_2$(*g*) | – | −135,060 + 23.43T |
| Si(*l*) | 0.0013 | −131,500 − 17.24T |
| Ti(*s*) | 0.038 | −31,129 − 44.98T |
| V(*s*) | 0.10 | −20,710 − 45.61T |
| W(*s*) | 1.20 | 31,380 − 63.60T |
| Zr(*s*) | 0.043 | −34,727 − 50.00T |

## 4.2 GAS SOLUBILITIES IN SOLID AND LIQUID IRON

### 4.2.1 SIEVERT'S LAW

Diatomic gases such as $O_2$, $S_2$, $N_2$ and $H_2$ dissolve in liquid and solid metals in the atomic form

$$\tfrac{1}{2}X_2(g) = [X] \qquad (4.6)$$

for which the isothermal equilibrium constant is

$$K = \frac{[\%X]}{(p_{X_2})^{1/2}} \qquad (4.7)$$

For ideal solutions, the concentration of X is directly proportional to the square root of the equilibrium gas partial pressure; this is known as the Sievert's law.

### 4.2.2 SOLUBILITY OF HYDROGEN

$$\tfrac{1}{2}H_2(g) = [H] \qquad\qquad (4.8)$$

$$K = \frac{[\text{ppm H}]}{(p_{H_2})^{1/2}} \qquad\qquad (4.9)$$

The temperature dependence of $K$ (with $p_{H_2}$ in atm) is as follows for $\alpha$, $\delta$(bcc) iron, $\gamma$(fcc) iron and liquid iron ($l$); see Fig. 4.5.

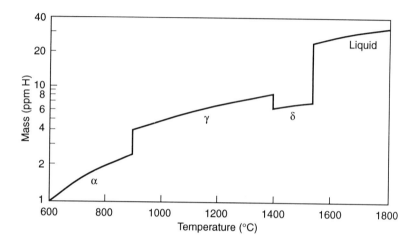

Fig. 4.5    Solubility of hydrogen in pure iron (or low-alloy steels) at 1 atm pressure of $H_2$.

$$\log K_{\alpha,\delta} = -\frac{1418}{T} + 1.628 \qquad\qquad (4.9a)$$

$$\log K_{\gamma} = -\frac{1182}{T} + 1.628 \qquad\qquad (4.9b)$$

$$\log K_{l} = -\frac{1900}{T} + 2.423 \qquad\qquad (4.9c)$$

where the temperature $T$ is in degrees Kelvin.

During rapid cooling of a heavy-section steel casting, e.g. thick slab or bloom, there will be little diffusion of H out of the casting. Because the hydrogen solubility decreases with a decreasing temperature, there will be a build up of $H_2$ pressure in the steel matrix during rapid cooling. For the limiting case of no hydrogen diffusion, the $H_2$ pressures will be as shown below in the steel containing 2, 4 and 8 ppm H.

| Temperature °C | | $H_2$ pressure, atm | | |
|---|---|---|---|---|
| | | 2 ppm H | 4 ppm H | 8 ppm H |
| γ | 1400 | 0.058 | 0.23 | 0.9 |
| | 1100 | 0.12 | 0.48 | 1.9 |
| | 900 | 0.23 | 0.92 | 3.7 |
| α | 900 | 0.58 | 2.33 | 9.3 |
| | 700 | 1.83 | 7.3 | 29.2 |
| | 500 | 10.4 | 41.6 | 166.5 |

Because of this pressure buildup, internal cracking and flaking could occur in heavy sections of crack sensitive steels, if the hydrogen content is too high. If degassing facilities are not available, heavy-section castings for critical applications are cooled slowly to allow the diffusion of hydrogen out of the steel casting, e.g. slab, bloom or ingot. Most steel works nowadays do have degassers. The HSLA grade steels for critical applications, e.g. armour plates, pressure vessels, rail steels etc., are degassed to remove hydrogen to 2 ppm or less before casting.

### 4.2.3 SOLUBILITY OF NITROGEN

The nitrogen solubility data are summarised by the following equations and are shown graphically in Fig. 4.6.

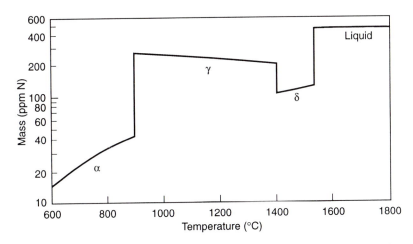

Fig. 4.6   Solubility of nitrogen in pure iron (or low-alloy steels) at 1 atm pressure of $N_2$.

$$\tfrac{1}{2}N_2 = [N] \tag{4.10}$$

$$K = \frac{[\text{ppm N}]}{(p_{N_2})^{1/2}} \tag{4.11}$$

$$\log K_{\alpha,\delta} = -\frac{1570}{T} + 2.98 \tag{4.11a}$$

$$\log K_{\gamma} = \frac{450}{T} + 2.05 \tag{4.11b}$$

$$\log K_{l} = -\frac{188}{T} + 2.76 \tag{4.11c}$$

### 4.2.4 SOLUBILITY OF CARBON MONOXIDE

Carbon monoxide dissolves in liquid iron (steel) by dissociating into atomic carbon and oxygen

$$CO(g) = [C] + [O] \tag{4.12}$$

for which the equilibrium constant (which is not too sensitive to temperature) is given by the following equation for low alloy steels containing less than 1% C.

$$K = \frac{[\%C][\text{ppm O}]}{p_{CO} \text{ (atm)}} = 20 \tag{4.13}$$

At higher carbon contents, a correction should be made to K with the activity coefficients $f_O^C$ ($\log f_O^C = -0.13[\%C]$) and $f_C^C$ ($\log f_C^C = 0.18[\%C]$).
    For the reaction

$$CO_2 (g) = CO (g) + [O] \tag{4.14}$$

$$K = \frac{p_{CO}}{p_{CO_2}} \quad [\text{ppm O}] = 1.1 \times 10^4 \text{ at } 1600°C \tag{4.15}$$

For 800 ppm O in low carbon steel at tap, the equilibrium ratio $p_{CO}/p_{CO_2}$ is 13.75. For this state of equilibrium, the gas mixture contains 6.8% $CO_2$ and 93.2% CO.

### 4.2.5 SOLUBILITY OF WATER VAPOUR

From the free energy of formation of water vapour and the solubilities of hydrogen and oxygen in liquid iron, the following equilibrium constant is obtained for the reaction of water vapour with liquid iron for 1600°C.

$$H_2O(g) = 2[H] + [O] \tag{4.16}$$

$$K = \frac{[\text{ppm H}]^2[\text{ppm O}]}{p_{H_2O}(\text{atm})} = 1.77 \times 10^6 \text{ at } 1600°C \tag{4.17}$$

The hydrogen and oxygen contents of low alloy liquid steel in equilibrium with $H_2$–$H_2O$ mixtures at 1 atm pressure and 1600°C are shown in Fig. 4.7.

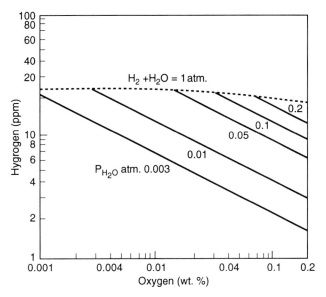

Fig. 4.7 Concentrations of hydrogen and oxygen in liquid iron at 1600°C in equilibrium with indicated compositions of $H_2$–$H_2O$ mixtures.

### 4.2.6 SOLUBILITY OF SULPHUR DIOXIDE

From the free energy of formation of $SO_2$ and the solubilities of gaseous sulphur and oxygen in liquid iron, the following equilibrium constant is obtained for the reaction of $SO_2$ with liquid iron at 1600°C.

$$SO_2(g) = [S] + 2[O] \qquad (4.18)$$

$$K = \frac{[\%S][\%O]^2}{p_{SO_2}(\text{atm})} = 1558 \text{ at } 1600°C \qquad (4.19)$$

For the concentrations of sulphur and oxygen present in liquid steel, it is seen that the corresponding equilibrium pressure of $SO_2$ is infinitesimally small. It is for this reason that no sulphur can be oxidised to $SO_2$ during steelmaking with oxygen blowing.

### 4.2.7 SOLUBILITY OF HYDROGEN SULPHIDE

$$H_2S(g) = [S] + H_2(g) \qquad (4.20)$$

$$K = \frac{p_{H_2}}{p_{H_2S}} \ [\text{ppm S}] = 4.3 \times 10^6 \text{ at } 1600°C \qquad (4.21)$$

For the steel containing 100 ppm S, the equilibrium gas pressure ratio is

$p_{H_2S}/p_{H_2}$ = 2.4 × 10$^{-5}$. This example illustrates that liquid steel cannot be desulphurised by purging with hydrogen, and that sulphur bearing gases in the reheating furnace atmosphere will result in sulphur pickup by the steel.

## 4.3 SELECTED IRON-BASE PHASE DIAGRAMS

### 4.3.1 CRYSTALLOGRAPHIC FORMS OF IRON

Iron occurs in two crystalline modifications as shown in Fig. 4.8: body-centred cubic (b.c.c) denoted by $\alpha$ and $\delta$ and face-centred cubic (f.c.c) denoted by $\gamma$. The phase transformations for pure iron occur at the following temperatures.

$$\alpha\text{–Fe} \xrightarrow{910°C} \gamma\text{–Fe} \xrightarrow{1392°C} \delta\text{–Fe} \xrightarrow{1537°C} \text{liquid Fe}$$

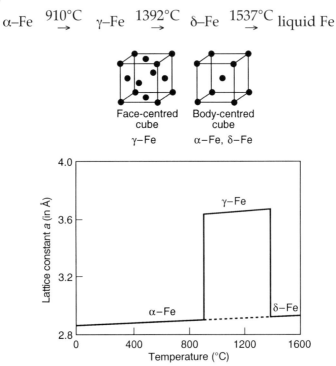

Fig. 4.8   Variation of lattice constant of pure iron with temperature.

In the b.c.c structure, the corner atoms of the cube become centre atoms in the body of the neighbouring cubic lattice, with every atom having eight nearest neighbours, i.e. coordination number eight. In the f.c.c structure, the corner atoms of the cube become centre atoms on the faces of the neighbouring cubic lattice, with coordination number twelve.

The interatomic distance, i.e. diameter $d$ of hypothetical tangent atom spheres, is related to the lattice parameter $a$ (length of the unit cube edge) as derived from the geometry of atom arrangement.

$$\text{for b.c.c.} \quad d = \frac{a\sqrt{3}}{2} \quad\quad (4.22a)$$

$$\text{for f.c.c.} \quad d = \frac{a}{\sqrt{2}} \quad\quad (4.22b)$$

The interatomic distances calculated from the lattice parameters at the phase transformation temperature 910°C (Fig. 4.8), are

| Phase | $a$, Å | $d$, Å* |
|-------|--------|---------|
| α | 2.898 | 2.510 |
| γ | 3.639 | 2.573 |

In the non-close-packed structures, the atoms behave as though they were slightly smaller than those in the close-packed structures.

In metallurgical language, α–Fe is called the ferrite phase and γ–Fe, the austenite phase. The ferrite phase goes through a magnetic transformation at 769°C, above which iron is not magnetic, however, this is not a true polymorphic transformation. Since the lattice constant of austenite is larger than that of ferrite, the elements forming interstitial solid solutions in iron, i.e. within the iron space lattice, have larger solubilities in the austenite phase, e.g. H, B, C, N and O. In the substitutional solid solutions, solute atoms replace the solvent atoms in the space lattice of the latter. Depending on the nature of the alloying elements in iron, the transformation temperatures vary over a wide range. The elements with f.c.c structure dissolved in iron favour the f.c.c. structure and therefore they extend the temperature stability range of the austenite phase, e.g. Ni, Co. Similarly a solute element with b.c.c structure extends the stability range of the ferrite phase, e.g. Si, Cr, V. The existence of these two crystalline forms of iron is largely responsible for the versatility of iron alloys in practical applications. The iron can be rendered malleable, ductile, tough or hard, by alloying with suitable elements and applying appropriate heat and mechanical treatments.

### 4.3.2 IRON-CARBON SYSTEM

Since carbon is one of the most important ingredients of steel, the study of the iron-carbon phase equilibrium diagram has received much attention during the past several decades. The phase diagram is given in Fig. 4.9.

*In SI units 10 Å ≡ nm (nanometre) = $10^{-9}$m.

There are three invariants in this system; peritectic at 1499°C, eutectic at 1152°C and eutectoid at 738°C. The phase boundaries shown by broken lines are for the metastable equilibrium of cementite, $Fe_3C$, with austenite. It was shown by Wells[3] that if sufficient time is allowed, iron-carbon alloys containing austenite and cementite decompose to austenite and graphite. The solubilities of graphite and of cementite in α-iron below the eutectoid temperature are given by the following equations using the data of Swartz[4]

$$\log \%C \text{ (graphite)} = - \frac{5250}{T} + 3.53 \qquad (4.23)$$

$$\log \%C \text{ (cementite)} = - \frac{3200}{T} + 1.50 \qquad (4.24)$$

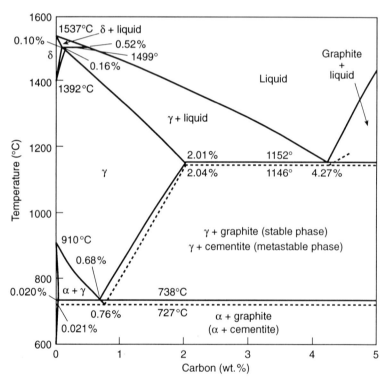

Fig. 4.9   Iron–carbon phase equilibrium diagram. Dashed lines represent phase boundary for metastable equilibrium with cementite.

The solubility of graphite in liquid iron is well established and the experimental data, reviewed by Neumann and Schenck[5], are summarised by the equation

$$[\%C] = 1.30 + 2.57 \times 10^{-3} \, T \, (°C) \qquad (4.25)$$

If atom fraction $N_C$ is used, the same set of data can be represented by the following equation in terms of $\log N_C$ and the reciprocal of the absolute temperature.

$$\log N_C = -\frac{560}{T} - 0.375 \tag{4.26}$$

The effect of silicon, phosphorus, sulphur, manganese, cobalt and nickel on the solubility of graphite in molten iron was determined by Turkdogan et al.[6,7] and graphite solubility in iron-silicon and iron-manganese melts by Chipman et al.[8,9]. Similar measurements with iron-chromium melts were made by Griffing and co-workers.[10] The experimental data are given graphically in Fig. 4.10 for 1500°C; the solubility at other temperatures can be estimated from this plot by using the temperature coefficient given in equation (4.25) for binary iron-carbon melts. In the iron–sulphur–carbon system there is a large miscibility gap. For example, at 1500°C the melt separates into two liquids containing

phase (I)    1.8% S and 4.24% C
phase (II) 26.5% S and 0.90% C

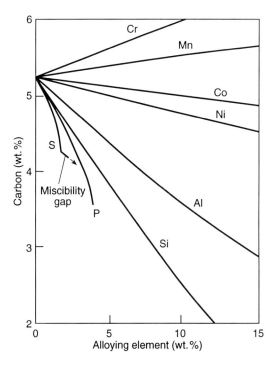

Fig. 4.10   Solubility of graphite in alloyed iron melts at 1500°C.

*4.3.2a Peritectic reaction*

The peritectic reaction occurring in the early stages of solidification of low-carbon steels is of particular importance in the continuous casting of steel. The peritectic region of the Fe–C system is shown on a larger scale in Fig. 4.11. As the temperature decreases within the two-phase region, δ + liquid, the carbon contents of δ-iron and residual liquid iron increase. At the peritectic temperature 1499°C, δ-iron containing 0.10% C reacts with liquid iron containing 0.52% C to form γ-iron with 0.16% C.

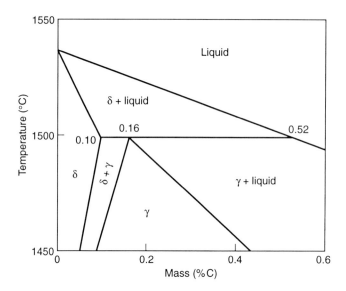

Fig. 4.11   Peritectic region of binary iron–carbon system.

The X-ray diffraction data give for the densities of iron-carbon alloys: 7.89 g cm⁻³ for δ–Fe with 0.10% C and 8.26 g cm⁻³ for γ–Fe with 0.16% C. Hence, the δ to γ phase transformation is accompanied by 4.7% volume shrinkage. Because of this shrinkage, the thin solidified shell in the mould of the caster will contract, producing a gap between the shell surface and mould wall. This situation leads to an uneven surface which is in partial contact with the mould wall, hence resulting in reduced heat flux at contracted areas. A reduced solidification growth rate and a nonuniform shell with thin spots, lowers the resistance of the steel to cracking which may cause a breakout in the mould. This phenomenon was well demonstrated experimentally by Singh and Blazek[11] using a bench scale caster; this subject is discussed further in a paper by Wolf and Kurz.[12]

In low alloy steels containing 0.10 to 0.16% C, the solid/liquid ratio at the peritectic invariant is higher than for steels containing more than 0.16% C. Therefore, due to the peritectic reaction, low alloy steels with 0.10

to 0.16% C are more susceptible to the development of surface cracks in continuous casting than steels with higher carbon contents.

### 4.3.2b Effect of alloying elements on peritectic invariant

As shown schematically in Fig. 4.12 for the ternary Fe–C–X, or multicomponent alloy steels, the peritectic reaction occurs over a temperature and composition range. From the experimental data on the liquidus and solidus temperatures of alloy steels, the carbon and temperature equivalents of alloying elements have been evaluated in three independent investigations.[13–15]

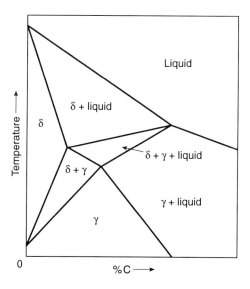

Fig. 4.12   Schematic representation of phase boundaries in low alloy steels as pseudo binary Fe–C system.

The carbon equivalents of the alloying elements for the peritectic reaction

$$C(\delta\text{–Fe}) + C(l\text{–Fe}) = C(\gamma\text{–Fe}) \qquad (4.27)$$

are usually formulated as follows

$$\text{Liquid phase:} \quad \Delta\%C = 0.52 + \Sigma\Delta C^X_{Pl} \cdot [\%X] \qquad (4.28)$$

$$\text{Delta phase:} \quad \Delta\%C = 0.10 + \Sigma\Delta C^X_{P\delta} \cdot [\%X] \qquad (4.29)$$

The changes in peritectic temperatures are formulated as

$$\text{Liquid phase:} \quad \Delta T = 1499°C + \Sigma\Delta T^X_{Pl} \cdot [\%X] \qquad (4.30)$$

$$\text{Delta phase:} \quad \Delta T = 1499°C + \Sigma\Delta T^X_{P\delta} \cdot [\%X] \qquad (4.31)$$

The coefficients $\Delta C_P^X$ and $\Delta T_P^X$ can be positive or negative, depending on the alloying element X.

If the reported values of these peritectic parameters are reliable numerically and by the + or − sign, they should correlate to one another in a consistent manner either empirically or theoretically. It is on the basis of such a criterion that a reassessment is now made of the values reported by Yamada *et al.*[14]

The slopes of the liquidus and solidus lines at low solute contents in the Fe–X binary systems are plotted in Fig. 4.13 against the peritectic temperature coefficients $\Delta T_{Pl}$ and $\Delta T_{P\delta}$ (°C/%X) in the Fe–C–X alloys. Of the elements considered here, Mn and Ni are the only two austenite stabilisers involving peritectic reactions, hence the peritectic temperature coefficients are positive.

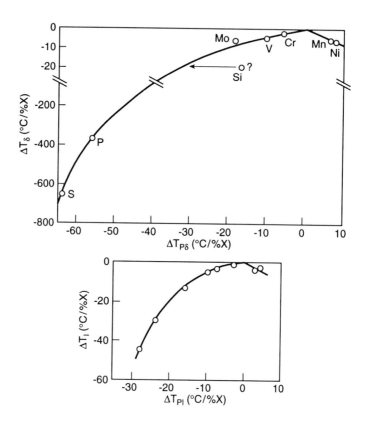

Fig. 4.13  Linear decrease in solidus, $\Delta T_\delta$, and liquidus, $\Delta T_l$, temperatures at low solute contents in binary systems Fe–X, related to changes in peritectic temperatures with respect to the binary Fe–C system.

As is seen from the plots in Fig. 4.14 the peritectic carbon coefficients vary in a systematic manner with the peritectic temperature coefficients. It appears that the reported value of $\Delta C_{P\delta} = 0.025$ for Mo may be lowered to 0.012.

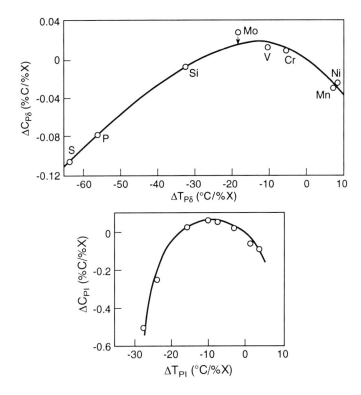

Fig. 4.14   Changes in peritectic carbon contents of delta and liquid iron related to changes in the respective peritectic temperatures with respect to the binary Fe–C system.

The correlations are shown in Fig. 4.15 between the peritectic tempera-ture and carbon coefficients. With the anticipation that the parameters for the delta and liquid phases may have the same sign, i.e. either positive or negative, the curves in Fig. 4.15 are drawn to pass through the origin of the coordinates. Suggested corrections in Fig. 4.15 to the values of Si and Mo are the same as those indicated in Figs. 4.13, 4.14.

There is no peritectic reaction in the Fe–X alloys with the ferrite stabilis-ing element X. Therefore, the addition of a ferrite stabilising element X to the Fe–C system should decrease the peritectic temperature coefficients. Also, the steeper the liquidus and solidus slopes in the Fe–X system, the greater will be the decrease in the peritectic temperature coefficients; this is evident from the experimental data in Fig. 4.15. The reported value of

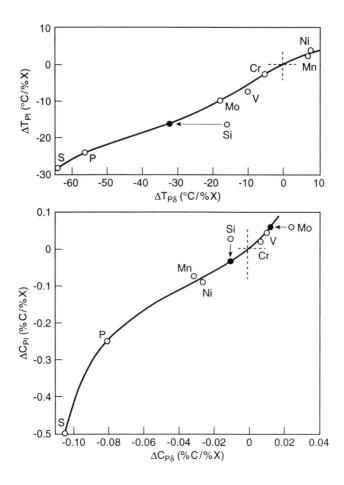

Fig. 4.15   Interrelations between peritectic temperature and composition changes.

$\Delta T_{P\delta}$ for Si appears to be low; to be consistent with the other data, we may take $\Delta T_{P\delta} = -32\ °C/\%Si$.

The peritectic temperature and carbon coefficients determined by Yamada *et al.* are listed in Table 4.4, with minor numerical adjustments to some of the parameters in accord with the empirical correlations given in Fig. 4.15. The composition ranges of the alloys used in the experimental work of Yamada *et al.* are indicated in Table 4.4. The peritectic parameters will have lower numerical values at higher concentrations of alloying elements. For continuous casting of the HSLA plate grade steels containing 1.0 to 1.3% Mn, the recommended carbon content is in the range 0.06 to 0.08% (max) to avoid the peritectic region, as indicated by $\Delta\%C/\%Mn = -0.029$ in Table 4.4.

**Table 4.4** Peritectic temperature and composition parameters in accord with the curves in Fig. 4.15

| Alloying Element | Composition range, wt.% | Δ°C/%X | | Δ%C/%X | |
|---|---|---|---|---|---|
| | | $\Delta T_{Pl}$ | $\Delta T_{P\delta}$ | $\Delta C_{Pl}$ | $\Delta C_{P\delta}$ |
| Cr | <1.5 | −3 | −5 | +0.022 | +0.006 |
| Mn | <1.5 | +3 | +7 | −0.085 | −0.029 |
| Mo | <1.5 | −10 | −18 | +0.055 | +0.012 |
| Ni | <3.5 | +4 | +9 | −0.082 | −0.027 |
| P | <0.05 | −24 | −56 | −0.250 | −0.080 |
| S | <0.03 | −28 | −64 | −0.500 | −0.105 |
| Si | <0.6 | −16 | −32 | −0.035 | −0.010 |
| V | <1.0 | −6 | −11 | +0.045 | +0.010 |

The phase boundaries in the peritectic region of the Fe–C–X alloys, projected on the Fe–C–X composition diagram, are shown schematically in Fig. 4.16 for alloys where X is: (a) ferrite stabiliser and (b) austenite stabiliser. In the Fe–C–Mn (or Ni) system, the peritectic reaction occurs at all compositions between the peritectic regions of the binaries Fe–C and Fe–Mn (Ni) systems. In the Fe–C alloys with ferrite stabilisers, the peritectic reaction will not occur beyond a certain concentration of the alloying element X as shown in FIg. 4.16a. This limiting case applies only to high alloy steels.

*4.3.2c Liquidus temperatures of low alloy steels*

The liquidus temperatures of low alloy steels are derived from the binary Fe–X systems on the assumption that the coefficients $\alpha = \Delta T/\%X$ are additive in their effects on the melting point of iron.

Below 0.5% C, where the solidification begins with the formation of delta (δ) iron, the following equation would apply

$$\text{Liquidus } T(°C) = 1537 - 73.1[\%C] + \Sigma\alpha[\%X] \qquad (4.32)$$

For the carbon contents within the range 0.5 to 1.0% C, where the solidification begins with the formation of gamma (γ) iron, the following equation is recommended.

$$\text{Liquidus } T(°C) = 1531 - 61.5[\%C] + \Sigma\alpha[\%X] \qquad (4.33)$$

The same coefficients α are used in both equations.

| Alloying element X | Coefficient α, °C/%X |
|---|---|
| Al | −2.5 |
| Cr | −1.5 |
| Mn | −4.0 |
| Mo | −5.0 |

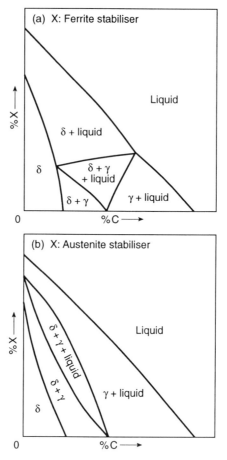

Fig. 4.16   Phase boundaries in the peritectic region of the Fe–C–X alloys projected on the composition diagram.

| Alloying element $X$ | Coefficient $\alpha$, °C/%X |
|---|---|
| Ni | −3.5 |
| P | −30.0 |
| Si | −14.0 |
| S | −45.0 |
| V | −4.0 |

4.3.3 IRON–OXYGEN SYSTEM

The thermodynamics of the iron–oxygen system was studied in detail by Darken and Gurry.[16] Fig. 4.17 gives the temperature-composition phase diagram for a total pressure of one atmosphere. There are two characteristic features of this system. Oxygen is soluble in iron to a limited extent only; at the eutectic temperature 1527°C, the maximum solubility is 0.16% O, above which a liquid oxide phase containing 22.6% O is formed. The second characteristic feature is the formation of wustite which has a variable composition and is not stable below 560°C. The stoichiometric ferrous oxide does not exist and wustite in equilibrium with iron has the composition corresponding to about $Fe_{0.95}O$ within the temperature range 800–1371°C (melting temperature). That is, wustite is deficient in iron cations and electroneutrality is maintained by the presence of some trivalent iron cations together with the divalent iron cations in wustite. Within the wustite phase, the ratio $Fe^{3+}/Fe^{2+}$ increases with an increasing partial pressure of oxygen. Although magnetite in equilibrium with wustite has the stoichiometric composition $Fe_3O_4$, in the presence of haematite the magnetite becomes deficient in iron with an increasing temperature.

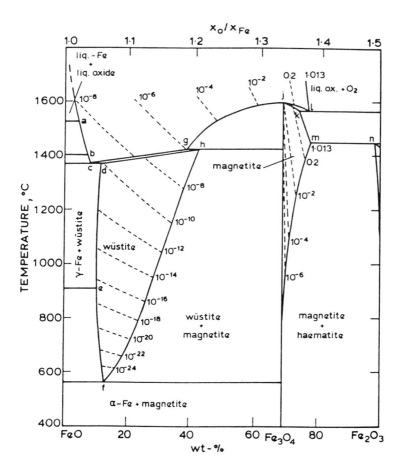

Fig. 4.17   Iron–oxygen phase equilibrium diagram between FeO and $Fe_2O_3$ compositions, based mainly on the data of Darken and Gurry[16]; broken lines in single phase regions are oxygen isobars in bar.

The solubility of oxygen in high-purity zone-refined iron was determined by Turkdogan *et al.*[17,18]; this was later confirmed by the work of Kusano *et al.*[19]. Iron side of the Fe–O phase diagram is shown in Fig. 4.18. At about 900°C, iron (α or γ) in equilibrium with wustite contains about 2 ppm O. At the peritectic invariant (1390°C) γ-iron containing 28 ppm O is in equilibrium with δ-iron and liquid iron oxide. In δ-iron equilibrated with liquid iron oxide, the oxygen content increases from 54 ppm at 1390°C to 82 ppm at the eutectic temperature 1527°C.

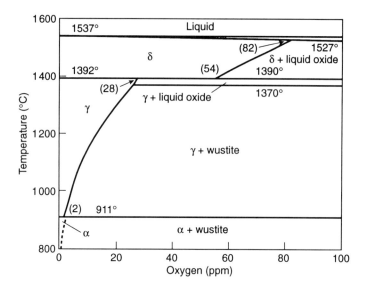

Fig. 4.18   Iron side of the iron–oxygen phase diagram. From Refs. 17,18.

Subsequent to earlier studies by various investigators, Taylor and Chipman[20] made the most reliable measurement of the oxygen solubility in liquid iron in equilibrium with essentially pure liquid iron oxide at temperatures of 1530–1700°C. In a later study, Distin *et al.*[21] extended the solubility measurements up to 1960°C. These two sets of experimental data plotted in Fig. 4.19 are represented by the equation

$$\log [\%0]_{sat.} = -\ \frac{6380}{T} + 2.765 \tag{4.34}$$

Numerous other solubility measurements made are in general accord with the data in Fig. 4.19. The oxygen content of liquid iron oxide in equilibrium with liquid iron decreases with an increasing temperature as given below and reaches the stoichiometric composition (22.27%) at about 2000°C.

| Temperature °C | %O in iron | %O in liquid iron oxide |
|---|---|---|
| 1527 (eutectic) | 0.16 | 22.60 |
| 1785 | 0.46 | 22.40 |
| 1880 | 0.63 | 22.37 |
| 1960 | 0.81 | 22.32 |

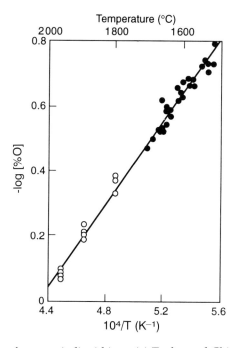

Fig. 4.19   Solubility of oxygen in liquid iron: (●) Taylor and Chipman, (○) Distin et al.

In a recent critical assessment of all the experimental thermochemical data on the iron–oxygen system, the author derived the following free energy equations.[22]

$$\tfrac{1}{2}O_2(g) = [O] \tag{4.35}$$

$$K = \frac{[\%O]f_O}{(p_{O_2})^{1/2}} \tag{4.35a}$$

at 1 mass % O, $\Delta G_O = -115{,}750 - 4.63T$ J (1800–2200K)    (4.35b)

where the activity coefficient $\log f_O = -0.10[\%O]$    (4.35c)

$$Fe(l) + [O] = FeO(l) \tag{4.36}$$

$$\Delta G° = -109{,}710 + 45.89T \text{ J (1800–2500K)} \tag{4.36a}$$

$$Fe(l) + \tfrac{1}{2}O_2(g) = FeO(l) \tag{4.37}$$

$$\Delta G° = -225{,}460 + 41.26T \text{ J (1800–2200K)} \tag{4.37a}$$

### 4.3.4 IRON–SULPHUR SYSTEM

The iron–sulphur phase equilibrium diagram is given in Fig. 4.20. The solubility of sulphur in solid iron was determined by Rosenqvist and Dunicz[23] and Turkdogan et al.[24]; the iron side of the iron-sulphur system thus evaluated is shown in Fig. 4.21. There is a peritectic invariant at 1365°C where the equilibrium phases are: $\gamma$-iron with 0.050 percent S, $\delta$-iron with 0.18 percent S and liquid containing 12 percent S. At the eutectic invariant 988°C, $\gamma$-iron in equilibrium with the liquid phase contains 0.012 percent S.

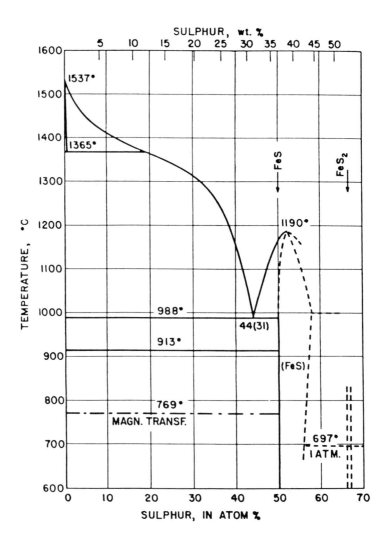

Fig. 4.20   Iron–sulphur phase equilibrium diagram.

4.3.5 IRON OXYSULPHIDE

Hot-shortness of steel is an old metallurgical problem that has been a subject of research for several decades. It has long been recognised that the presence of oxygen and/or sulphur in steel is the primary cause of hot-shortness, which usually occurs within the temperature range 900 to 1100°C. The hot-short range of course varies with the type of steel, particularly with the %Mn/%S ratio in the steel. From the phase diagram (Fig. 4.21) it is predicted that pure iron-sulphur alloys containing less than 100 ppm S should not be subject to hot-shortness during the entire hot-working operation because no liquid sulphide can form. On the other hand, an iron containing 200 ppm S will produce a small quantity of a liquid phase as its temperature falls below 1100°C. In the presence of oxygen, a liquid oxysulphide will form in γ-iron at lower sulphur contents and temperatures. As shown in Fig. 4.22, the sulphur content of iron in equilibrium with wustite and liquid oxysulphide reaches a maximum of 143 ppm at 1200°C; this is about one half of that in purified iron. At the ternary eutectic invariant (913°C) the sulphur in solution in equilibrium with wustite and liquid oxysulphide is 70 ppm.

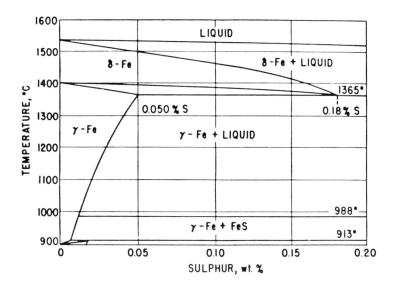

Fig. 4.21   Iron side of the Fe–S phase diagram. From Refs. 23,24.

There is an extensive liquid miscibility gap in the system Fe–S–O, as shown in Fig. 4.23 based on the work of Hilty and Crafts[26] and of Bog and Rosenqvist.[27] The isothermal sections in Fig. 4.24 show the phase equilibria at 1100 and 1400°C.

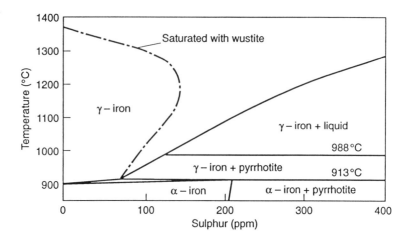

Fig. 4.22   Sulphur content of γ–Fe saturated with wustite is superimposed in binary Fe–S phase diagram. From Ref. 25.

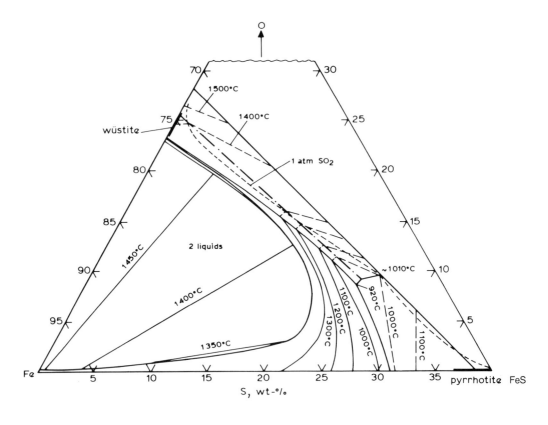

Fig. 4.23   Projection of the miscibility gap boundary and the liquidus isotherms on the composition diagram for the system Fe–S–O. From Refs. 26 and 27.

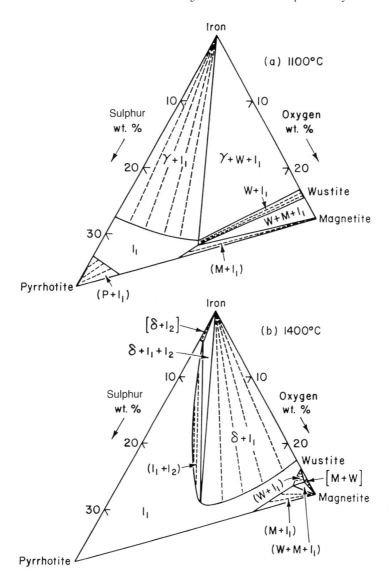

Fig. 4.24 1100° and 1400°C isotherms for Fe–S–O system; $\gamma$, gamma iron; $W$, wustite; $P$, pyrrhotite; $l_1$, liquid oxysulphide; $l_2$, liquid metal.

4.3.6 IRON–MANGANESE OXYSULPHIDE

The addition of manganese to steel is a well-known remedy to hot-shortness by virtue of forming solid oxides and sulphides. The solubility of MnS in $\gamma$-iron determined by Turkdogan *et al.*[24] is represented by

$$MnS(s) = [Mn] + [S]$$

$$\log [\%Mn][\%S] = -\frac{9020}{T} + 2.93 \qquad (4.38)$$

The univariant equilibria in the Fe–Mn–S–O system in the presence of γ-iron and Mn(Fe)O phases, derived by Turkdogan and Kor[25] are shown in Fig. 4.25. For manganese contents in the region above the univariant *j*, no liquid phase is present; below this curve, liquid oxysulphide forms. That is, as long as the steel contains Mn(Fe)O and Mn(Fe)S in equilibrium with iron, a liquid oxysulphide may form somewhere between 900° and 1225°C, depending on the concentration of manganese in solution in steel (in the absence of other alloying elements), e.g. for 10 ppm Mn 900°C and for ~ 10 percent Mn, ~ 1220°C. In the absence of oxygen however, a very small amount of manganese (e.g. 20 ppm Mn at 1000°C and 400 ppm at ~ 1300°C) is sufficient to suppress the formation of liquid sulphide.

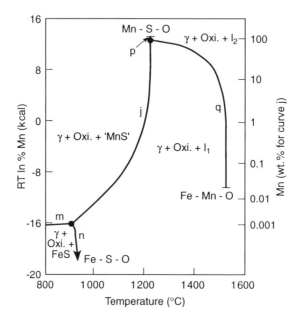

Fig. 4.25   Univariant equilibria in Fe–Mn–S–O system in the presence of γ–Fe and Mn(Fe)O phases. From Ref. 25.

j) γ–Fe, Oxi., 'MnS', $l_1$
m) γ–Fe, Oxi., 'MnS', 'FeS'
n) γ–Fe, Oxi., 'FeS', $l_1$
p) γ–Fe, Oxi., 'MnS', $l_2$
q) γ–Fe, Oxi., $l_1$, $l_2$

4.3.7 ELEMENTS OF LOW SOLUBILITY IN LIQUID IRON

A few elements of low solubility in liquid iron play some role in the steelmaking technology; a brief comment on the chemistry of such elements in liquid iron is considered desirable.

(a) Lead:

The break out of the furnace lining is often blamed on the presence of lead in the melt. Small amounts of metal trapped in the crevices of the lining are likely to get oxidised subsequent to tapping. The lead oxide together with iron oxide will readily flux the furnace lining, hence widening the cracks which ultimately leads to failure of the lining. The solubility of lead in liquid iron is sufficiently high that in normal steelmaking practice there should be no accumulation of lead at the bottom of the melt, except perhaps in the early stages of the melting of lead-containing scrap.

At steelmaking temperatures the vapour pressure of lead is about 0.5 atm and the solubility[28] in liquid iron is about 0.24 percent Pb at 1500°C increasing to about 0.4 percent Pb at 1700°C. The free-machining leaded steels contain 0.15 to 0.35 percent Pb. Evidently lead added to such steels is in solution in the metal prior to casting and precipitates as small lead spheroids during the early stages of freezing.

(b) Calcium:

The boiling point of calcium is 1500°C and its solubility in liquid iron is very low. Sponseller and Flinn[29] measured the solubility of calcium in iron at 1607°C for which the calcium vapour pressure is 1.69 atm. Under these conditions at 1607°C, 0.032 percent Ca is in solution. That is, the solubility at 1 atm pressure is 189 ppm Ca. They also investigated the effect of some alloying elements on the solubility; as seen from the data in Fig. 4.26, C, Si, Ni and Al increase markedly the calcium solubility in liquid iron. In melts saturated with $CaC_2$ the solubility of calcium of course decreases with an increasing carbon content.

(c) Magnesium:

The solubility of magnesium in iron-carbon alloys determined by Trojan and Flinn[30], is shown in Fig. 4.27 as functions of temperature and carbon content. Subsequently, Guichelaar *et al.*[31], made similar measurements with liquid Fe–Si–Mg alloys. Their data have been used in numerous studies to derive the equilibrium relations for the solubility of magnesium in liquid iron. However, there are some variations in the interpretation of the above mentioned experimental data. A reassessment of these experimental data is considered desirable.

The equilibrium relation for the solubility of Mg (in units of mass % $atm^{-1}$) is represented by

$$Mg(g) = [Mg]$$

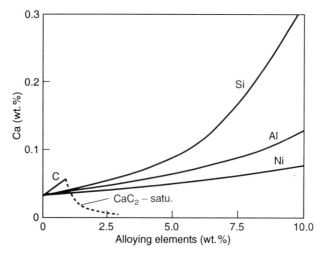

Fig. 4.26   Effect of alloying elements on the solubility of liquid calcium in liquid iron at 1607°C, corresponding to 1.69 atm pressure of calcium vapour. From Ref. 29.

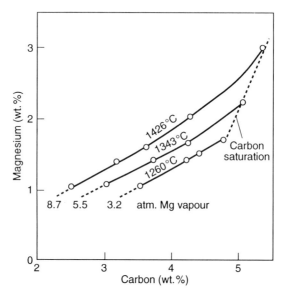

Fig. 4.27   Solubility of magnesium in liquid iron–carbon alloys at indicated temperatures and pressures of magnesium vapour. From Ref. 30.

$$K_{Mg} = \frac{[\%Mg]f_{Mg}}{P_{Mg}} \qquad (4.39)$$

where $f_{Mg}$ is the activity coefficient affected by the alloying elements. In the experiments with the Fe–C–Mg melts coexistent with liquid Mg, the latter contained less than 2 percent Fe, therefore, the Mg vapour pressure prevailing in the reactor would be essentially the same as that for pure Mg for which the following is obtained from the data of Guichelaar, *et al.*

$$\log P^\circ{}_{Mg}(atm) = -\frac{6730}{T} + 4.94 \qquad (4.40)$$

In the experiments with the Fe–Si–Mg alloys, the Mg-rich phase contained relatively large concentrations of Fe and Si. In addition to determining the miscibility gap in this system, Guichelaar *et al.*, also measured the equilibrium vapour pressure of Mg in the system by the 'boiling method.'

These experimental data are presented in Fig. 4.28 as a plot of log ([%Mg]/$p_{Mg}$) versus the concentration of C or Si in the iron-rich phase. For the Fe–C–Mg system, there is a linear relation; for the Fe–Si–Mg system, a linear relation holds up to about 18 percent Si. From the slopes of the lines, the following interaction coefficients are obtained:

$$e^C_{Mg} = -0.15 \text{ and } e\,^{Si}_{Mg} = -0.046$$

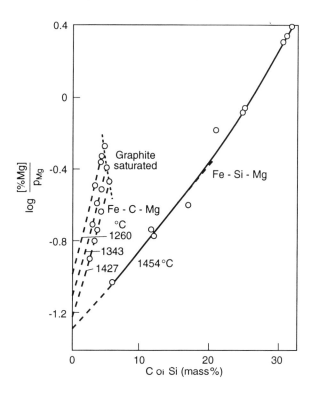

Fig. 4.28   Solubility of Mg in liquid Fe–C and Fe–Si alloys, derived from data in Refs. 30 and 31.

The intercepts of the lines with the ordinate axis at %C and %Si → 0 give the equilibrium constant $K_{Mg}$ for pure liquid iron (undercooled). The temperature dependence of $K_{Mg}$ shown in Fig. 4.29 is represented by

$$\log K_{Mg} = \frac{4110}{T} - 3.698 \tag{4.41}$$

which gives the following free energy equation for the solution of Mg vapour in liquid iron:

$$Mg(g) = [Mg]\,(1\%) \tag{4.42}$$

$$\Delta G_S = -78,690 + 70.80\,T\ \text{J mol}^{-1} \tag{4.42a}$$

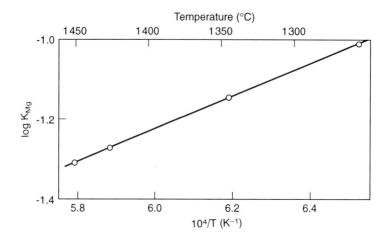

Fig. 4.29   Temperature dependence of the quilibrium constant for the solubility of Mg in liquid iron (undercooled), %Mg atm$^{-1}$.

For the quaternary melts Fe–C–Si–Mg, the activity coefficient of Mg may be approximated by the product

$$f_{Mg} = f_{Mg}^{C} \times f_{Mg}^{Si} \tag{4.43}$$

The solubility of Mg in liquid Fe–C–Si–Mg alloys calculated from the above equations agrees well with the measured values reported by Trojan and Flinn.

In the experiments by Engh et al.[32] the Mg solubility in graphite-saturated liquid iron at 1260°C was measured by equilibrating the liquid Fe–C–Mg alloys with liquid Pb-Mg alloys. The vapour pressure of Mg in the two-liquid system was derived from equation (4.40) and the known activity coefficient of Mg in the Pb-Mg phase. The ratio %Mg/$p_{Mg}$(atm) = 0.553 they obtained for graphite-saturated iron compares well with 0.525 determined by Trojan and Flinn. Noticeably higher values were reported by other investigators: 0.80 by Speer and Parlee[33] and 0.70 by Irons and Guthrie.[34]

## 4.4 SOLUBILITIES OF CaO, CaS, MgO AND MgS IN LIQUID IRON

There are large variations in the experimental data for the solubilities of oxides and sulphides of calcium and magnesium in liquid iron as evident from the data listed in Table 4.5 for 1600°C. For low concentrations of Ca, Mg, O and S in liquid iron, their activity coefficients $f_i$ will be essentially unity, therefore the solubility products for unit activities of oxides and sulphides are given in terms of mass percentages of the solutes.

**Table 4.5**   Experimentally determined solubility products of CaO, MgO, CaS and MgS are compared with those calculated from the thermochemical data for 1600°C.

| Solubility product | Experimental | Ref. | Calculated |
|---|---|---|---|
| $K_{CaO} = [\%Ca][\%O]$ | $9.0 \times 10^{-7}$ | 37 | $6.0 \times 10^{-10}$ |
| | $1.0 \times 10^{-10}$ | 38 | |
| | $2.5 \times 10^{-9}$ | 39 | |
| | $4.5 \times 10^{-8}$ | 40 | |
| | $5.5 \times 10^{-9}$ | 41 | |
| $K_{MgO} = [\%Mg][\%O]$ | $2.0 \times 10^{-6}$ | 37 | $2.4 \times 10^{-9}$ |
| | $7.5 \times 10^{-6}$ | 42 | |
| $K_{CaS} = [\%Ca][\%S]$ | $1.7 \times 10^{-5}$ | 37 | $2.0 \times 10^{-9}$ |
| | $7.9 \times 10^{-8}$ | 39 | |
| | $3.2 \times 10^{-7}$ | 40 | |
| $K_{MgS} = [\%Mg][\%S]$ | $5.9 \times 10^{-2}$ | 37 | $1.2 \times 10^{-4}$ |

The solubility products measured by Nadif and Gatellier[37] are much higher than those of other investigators and several orders of magnitude greater than those calculated from the free energy data and solubilities of Ca($g$) and Mg($g$) in liquid iron already discussed.

It should be noted that the free energy data used for CaO and MgO are those determined recently by Gourishankar *et al.*[35] using the Knudsen effusion technique. They found that for the reactions

$$Ca(g) + \tfrac{1}{2}O_2(g) = CaO(s) \qquad (4.44)$$

$$Mg(g) + \tfrac{1}{2}O_2(g) = MgO(s) \qquad (4.45)$$

the standard free energy data given in JANAF Tables,[36] are in error and should be corrected as follows:

$$\Delta G_T^\circ = \Delta G_T^\circ \text{ (tabulated)} + 33 \text{ kJ mol}^{-1} \text{ CaO} \qquad (4.44a)$$

$$\Delta G_T^\circ = \Delta G_T^\circ \text{ (tabulated)} - 34 \text{ kJ mol}^{-1} \text{ MgO} \qquad (4.45a)$$

The experimental data of Gourishankar *et al.* for reactions (4.44) and (4.45) may be represented by the following free energy equations.

For CaO:    $\Delta G_T^\circ = -753{,}880 + 192.5T$ J (1900–2100 K)    (4.46)

For MgO:    $\Delta G_T^\circ = -759{,}380 + 202.6T$ J (1800–2100 K)    (4.47)

These equations are used in the free energy data given in Table 1.1 of Chapter 1.

For the reaction equilibrium

$$MgS(s) = [Mg] + [S] \tag{4.48}$$

$$K = [\%Mg][\%S] \quad f_{Mg}^C \ f_S^C \tag{4.48a}$$

$$\log K = -\frac{17{,}026}{T} + 5.161 \tag{4.48b}$$

where the activity coefficients $f_{Mg}^C$ and $f_S^C \rightarrow 1$ for low concentrations of Mg, S and C dissolved in iron. For graphite-saturated liquid iron at 1260°C with 4.4% C, for which $f_{Mg}^C = 0.22$ and $f_S^C = 4.4$, the calculated solubility product is $[\%Mg][\%S] = 1.2 \times 10^{-6}$; this agrees well with the experimental data of Engh et al.[32] plotted in Fig. 4.30. Some of the data points lying above the calculated line are indicative of the presence of a small quantity of dispersed particles of MgS in the melt; the same point of view was expressed by Engh et al. in their paper.

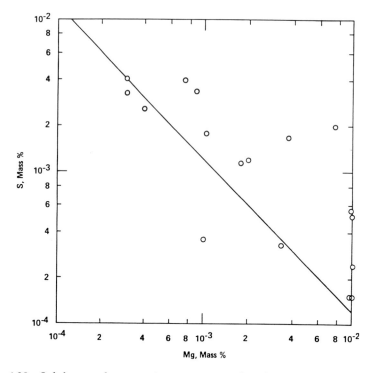

Fig. 4.30  Sulphur and magnesium contents of carbon-saturated liquid iron at 1260°C. From Ref. 32. The line is calculated for MgS saturation, i.e. $[\%Mg][\%S] = 1.2 \times 10^{-6}$.

### 4.5 SOLUBILITIES OF NITRIDES AND CARBIDES IN SOLID AND LIQUID IRON

For steel compositions of practical interest, the nitrides and carbides do not form in liquid steel. These compounds precipitate during cooling of the solidified steel along the austenite grain boundaries. These precipitates markedly influence the mechanical properties of steel both at room temperature and hot working temperatures.

Representing the nitride and carbide by $XY_n$, the reaction to be considered is

$$[X] + n[Y] = XY_n \qquad (4.49)$$

where $Y$ is nitrogen or carbon, and $X$ the alloying element in steel. The isothermal state of equilibrium is represented by the solubility product

$$K = [\%X][\%Y]^n \qquad (4.50)$$

where the temperature dependence of the equilibrium constant $K$ is of the form

$$\log K = - \frac{A}{T} + B \qquad (4.51)$$

For the elements dissolved in steel, when the product $[\%X] [\%Y]^n$ exceeds the equilibrium solubility product $K$, the nitride or carbide will precipitate, if the reaction kinetics are favourable.

Example:
Calculate the temperature at which AlN and Nb(C,N) begins to precipitate under equilibrium conditions in steel containing in solution 0.02% Al, 0.04% Nb, 0.08% C and 0.006% N.
For AlN:

$$\log [0.02 \times 0.006] = - \frac{6770}{T} + 1.03 = - 3.921$$

$$\therefore T = \frac{6770}{4.951} = 1367°K \equiv 1094°C$$

For $NbC_{0.7}N_{0.2}$:

$$\log [0.04 \times 0.08^{0.7} \times 0.006^{0.2}] = - \frac{9450}{T} + 4.12 = - 2.61$$

$$\therefore T = \frac{9450}{6.73} = 1404°K \equiv 1131°C$$

If there is equilibrium precipitation of AlN and Nb(C,N) in homogenised austenite, the concentrations of Al, Nb and N in solution will decrease with a decreasing temperature as shown in Fig. 4.31a. On the other hand, if Al and Nb are concentrated at the austenite grain boundaries as a consequence of solute microsegregation during solidification, the nitride and carbonitride precipitation under equilibrium conditions will begin at higher temperatures, as shown in Fig. 4.31b. This subject will be discussed in more detail in Chapter 10.

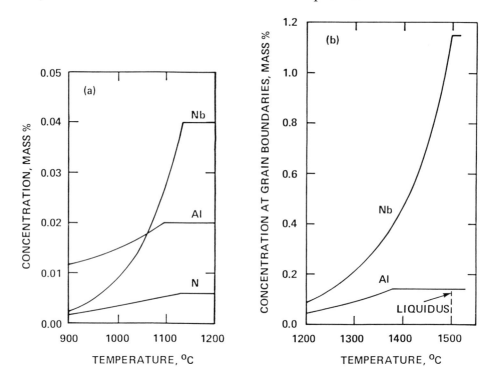

Fig. 4.31 Equilibrium precipitation of AlN and Nb (C,N) in austenite during cooling: (a) for homogenous solute distribution in steel, (b) for solutes segregated to austenite grain boundaries.

The reassessed values of the solubility products of carbides and nitrides in ferrite, austenite and liquid iron are given in Table 4.6, reproduced from the author's previous publication.[43] The temperature dependence of the solubility product $K_\gamma$ in austenite is shown in Fig. 4.32.

**4.6 SURFACE TENSION OF LIQUID IRON ALLOYS**

The experimental data for the surface tension of liquid iron and its binary alloys have been compiled recently by Keene.[44] For purified liquid iron, the average value of the surface tension at temperature $T(°C)$ is represented by

$$\sigma_{Fe} = (2367 \pm 500) - 0.34T(°C), \text{ mNm}^{-1}$$

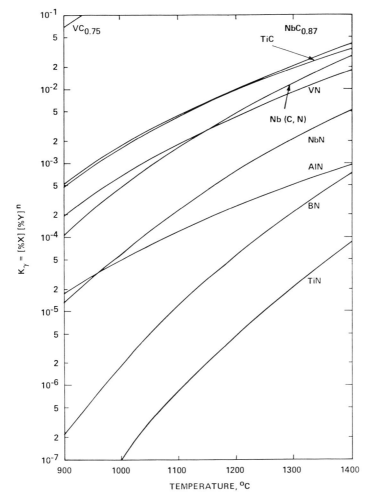

Fig. 4.32 Equilibrium solubility products of carbides and nitrides in austenite.

Keene derived the following weighted average limiting values of $\sigma$ (mNm$^{-1}$) for dilute solutions of $X$ in Fe–$X$ binary alloys.

Fe–C:   Virtually no effect of C on $\sigma_{Fe}$

Fe–Ce:   $\sigma = \sigma_{Fe} - 700[\%Ce]$

Fe–Mn:   $\sigma = \sigma_{Fe} - 51[\%Mn]$; $\partial\sigma/\partial T = -0.22$

Fe–N:   $\sigma = \sigma_{Fe} - 5585[\%N]$

Fe–P:   $\sigma = \sigma_{Fe} - 25[\%P]$

Fe–S:   (Discussed later)

Fe–Si:   $\sigma = \sigma_{Fe} - 30[\%Si]$; $\partial\sigma/\partial T = -0.25$

Gibbs' exact treatment of surface thermodynamics gives, for fixed unit surface area and constant temperature and pressure,

**Table 4.6**   Reassessed values of the solubility products of carbides and nitrides in iron

| Solubility product | $\log K_\gamma$ | $\log K_\alpha$ | $\log K_l$ |
|---|---|---|---|
| [%Al][%N] | $-\dfrac{6{,}770}{T} + 1.03$ | $-\dfrac{8{,}790}{T} + 2.05$ | $-\dfrac{12{,}950}{T} + 5.58$ |
| [%B][%N] | $-\dfrac{13{,}970}{T} + 5.24$ | $-\dfrac{14{,}250}{T} + 4.61$ | $-\dfrac{10{,}030}{T} + 4.64$ |
| [%Nb][%N] | $-\dfrac{10{,}150}{T} + 3.79$ | $-\dfrac{12{,}170}{T} + 4.91$ | |
| [%Nb][%C]$^{0.87}$ | $-\dfrac{7{,}020}{T} + 2.81$ | $-\dfrac{9{,}830}{T} + 4.33$ | |
| [%Nb][%C]$^{07}$[%N]$^{0.2}$ | $-\dfrac{9{,}450}{T} + 4.12$ | $-\dfrac{12{,}120}{T} + 5.57$ | |
| [%Ti][%N] | $-\dfrac{15{,}790}{T} + 5.40$ | $-\dfrac{18{,}420}{T} + 6.40$ | $-\dfrac{17{,}040}{T} + 6.40$ |
| [%Ti][%C] | $-\dfrac{7{,}000}{T} + 2.75$ | $-\dfrac{10{,}230}{T} + 4.45$ | $-\dfrac{6{,}160}{T} + 3.25$ |
| [%V][%N] | $-\dfrac{7{,}700}{T} + 2.86$ | $-\dfrac{9{,}720}{T} + 3.90$ | |
| [%V][%C]$^{0.75}$ | $-\dfrac{6{,}560}{T} + 4.45$ | $-\dfrac{7{,}050}{T} + 4.24$ | |

$$d\sigma = - RT \sum_i^k \Gamma_i \, d(\ln a_i) \qquad (4.52)$$

where $\Gamma_i$ is the surface excess concentration of the $i$th component and $a_i$ its activity. For a ternary system, equation (4.52) is reduced to

$$d\sigma = - RT \left( \Gamma_2 d (\ln a_2) + \Gamma_3 d (\ln a_3) \right) \qquad (4.53)$$

Since carbon dissolved in iron has virtually no effect on the surface tension of liquid iron, for the ternary system Fe–C–S, equation (4.53) is simplified to

$$d\sigma = - RT \, \Gamma_S d(\ln a_S) \qquad (4.54)$$

As shown, for example by Kozakevitch,[45] the addition of carbon to iron-sulphur alloys lowers the surface tension (Fig. 4.33). In the lower diagram, the plot of surface tension against the activity of sulphur illustrates that the apparent effect of carbon on the surface tension results from the effect of carbon on the activity coefficient of sulphur dissolved in iron.

As is seen from the compiled data in Fig. 4.34,* the surface tension data of various investigators for the Fe–S melts are in close agreement. The points read off from the curves in Figs. 4.33 and 4.34 are plotted in Fig. 4.35 as $\sigma$ versus $\log a_S$. It is seen that for $a_S > 0.01$, $\sigma$ is a linear function of $\log a_S$,

*References to data in Fig. 4.34 are given in a previous publication by the author.[46]

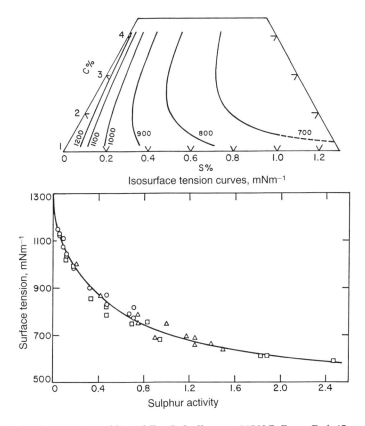

Fig. 4.33   Surface tension of liquid Fe–C–S alloys at 1450°C. From Ref. 45.

a limiting case for almost complete surface coverage with chemisorbed S. The shaded area represents the data of Selcuk and Kirkwood[47] for 1200°C.

## 4.7 DENSITY OF LIQUID STEEL

The temperature dependence of the densities of liquid iron, nickel, cobalt, copper, chromium, manganese, vanadium and titanium are given by the following equations as a linear function of temperature in °C in gcm$^{-3}$:

| | |
|---|---|
| Iron: | $8.30 - 8.36 \times 10^{-4}T$ |
| Nickel: | $9.60 - 12.00 \times 10^{-4}T$ |
| Cobalt: | $9.57 - 10.17 \times 10^{-4}T$ |
| Copper: | $9.11 - 9.44 \times 10^{-4}T$ |
| Chromium: | $7.83 - 7.23 \times 10^{-4}T$ |
| Manganese: | $7.17 - 9.30 \times 10^{-4}T$ |
| Vanadium: | $6.06 - 3.20 \times 10^{-4}T$ |
| Titanium: | $4.58 - 2.26 \times 10^{-4}T$ |

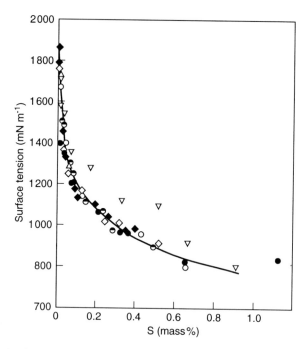

Fig. 4.34   Surface tension of Fe–S alloys at 1550–1600°C.

The specific volume and density of liquid iron-carbon alloys are given in Fig. 4.36 for various temperatures. It should be noted that the density of the liquid in equilibrium with austenite does not change much over the entire liquidus range.

## 4.8 Viscosity

The viscosity is a measure of resistance of the fluid to flow when subjected to an external force. As conceived by Newton, the shear stress $\varepsilon$, i.e. force per unit area, causing a relative motion of two adjacent layers in a fluid is proportional to the velocity gradient $du/dz$, normal to the direction of the applied force

$$\varepsilon = \eta \frac{du}{dz}$$

where the proportionality factor $\eta$ is viscosity of the fluid (liquid or gas).
    Viscosity unit: poise (g cm$^{-1}$s$^{-1}$) $\equiv$ 0.1N s m$^{-2}$.
    From theoretical considerations and experimental data, Andrade[49] derived the following relation for viscosities of liquid metals at their melting temperatures.

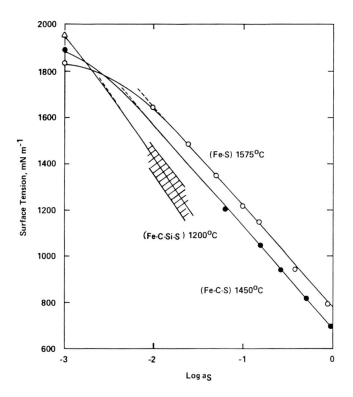

Fig. 4.35   Surface tension of Fe–S, Fe–C–S and Fe–C–Si–S alloys related to sulphur activity; $a_S \to$ %S when %C and %S $\to$ 0.

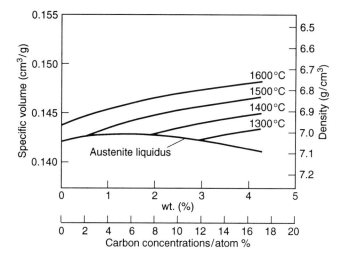

Fig. 4.36   Density of iron–carbon alloys. From Ref. 48.

$$\eta = 5.4 \times 10^{-4}(MT_m)^{1/2} V^{-2/3} \tag{4.55}$$

where $M$ is the atomic mass in $g$ and $V = M/\rho$ the molar volume.

The coefficients of viscosity of iron, cobalt, nickel and copper measured by Cavalier[50] are given in Fig. 4.37. In these experiments Cavalier was able to supercool metals 140°–170°C below their melting points and as seen from the data, there are no discontinuities in the viscosity coefficient vs temperature lines extending below the melting temperatures of the metals. The viscosities of Fe–C alloys determined by Barfield and Kitchener[51] are given in Fig. 4.38. In the iron-carbon melts the coefficient of viscosity is essentially independent of composition within the range 0.8 to 2.5 percent C; above 2.5 percent C, the viscosity decreases continuously with an increasing carbon content.

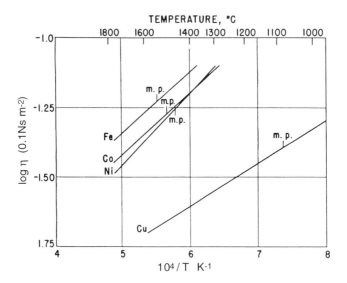

Fig. 4.37   Temperature dependence of viscosity coefficient of liquid iron, cobalt, nickel and copper. From Ref. 50.

### 4.9 MASS DIFFUSIVITIES IN SOLID AND LIQUID IRON

The diffusivity is an exponential function of temperature,

$$D = D_o \exp(-E/RT) \tag{4.56}$$

where $D_o$ is a constant for a given solute and E the activation energy (enthalpy) for the diffusion process.

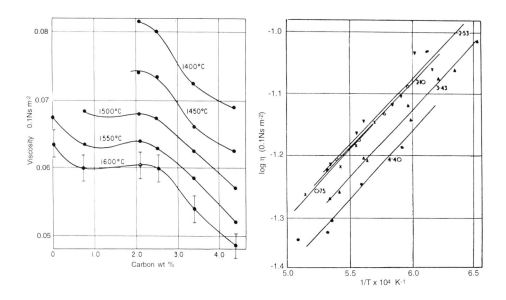

Fig. 4.38 Composition and temperature dependence of the viscosity coefficient of iron–carbon melts. From Ref. 51.

The general trend in the temperature dependence of solute diffusivity in solid iron is shown in Fig. 4.39. The interstitial elements, e.g. O, N, C, B, H have diffusivities much greater than the substitutional elements. Because of larger interatomic spacing, the diffusivities of interstitials in bcc-iron are greater and the heats of activations (~85kJ) are smaller than those in the fcc-iron with $E \sim 170$kJ. With the substitutional elements, the heat of activation is within 210 to 250 kJ for bcc-iron and within 250 to 290 kJ for fcc-iron.

In liquid iron alloys, diffusivities of elements are within $10^{-5}$ to $10^{-4}$ cm$^2$s$^{-1}$ with $E$ within 15 to 50 kJ.

Some solute diffusivity data for liquid iron and iron-carbon alloys are given in Table 4.7.

## 4.10 ELECTRICAL CONDUCTIVITY

The electrical conductivity $\lambda$ in the units of $\Omega^{-1}$cm$^{-1}$ is the reciprocal of the electrical resistivity. The electrical conductivity of liquid low alloy steel is about $\lambda = 7140 \; \Omega^{-1}$cm$^{-1}$ at steelmaking temperatures.

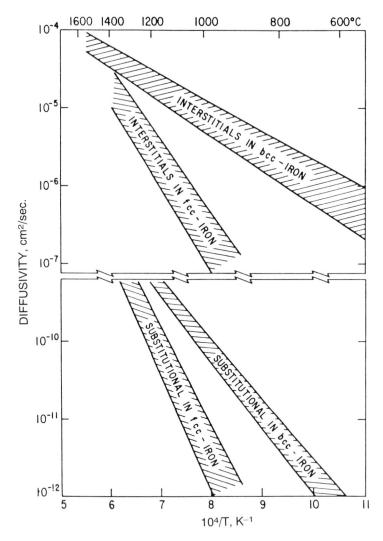

Fig. 4.39   Range of diffusivities of interstitial and substitutional elements in bcc and fcc iron.

## 4.11 THERMAL CONDUCTIVITY

From Fick's law, the thermal conductivity is defined by

$$\kappa = - \frac{Q}{\partial T / \partial x} \qquad (4.57)$$

where $Q$ is heat flux, energy per unit area per unit time, $\partial T / \partial x$ the temperature gradient normal to the direction of heat flow and $\kappa$ the thermal conductivity.

**Table 4.7** Selected solute diffusivities in liquid Fe–C alloys

| Diffusing element | Concentration mass % | Medium | Temp. range °C | $D$ cm²sec⁻¹ | $D_o$ cm²sec⁻¹ | E kJ |
|---|---|---|---|---|---|---|
| C | 0.03 | Fe | 1550 | $7.9 \times 10^{-5}$ | – | – |
| C | 2.1 | Fe | 1550 | $7.8 \times 10^{-5}$ | – | – |
| C | 3.5 | Fe | 1550 | $6.7 \times 10^{-5}$ | – | – |
| Co | Dilute sol. | Fe | 1568 | $4.7 \times 10^{-5}$ | – | – |
| Co | Dilute sol. | Fe | 1638 | $5.3 \times 10^{-5}$ | – | – |
| Fe | – | Fe–4.6%C | 1240–1360 | – | $4.3 \times 10^{-3}$ | 51 |
| Fe | – | Fe–2.5%C | 1340–1400 | – | $1.0 \times 10^{-2}$ | 66 |
| H | Dilute sol. | Fe | 1565–1679 | – | $3.2 \times 10^{-3}$ | 14 |
| Mn | 2.5 | gr.satu.Fe | 1300–1600 | – | $1.93 \times 10^{-4}$ | 24 |
| N | Dilute sol. | Fe | 1600 | $1.1 \times 10^{-4}$ | – | – |
| N | Dilute sol. | Fe–0.15%C | 1600 | $5.6 \times 10^{-5}$ | – | – |
| O | Dilute sol. | Fe | 1600 | $5.0 \times 10^{-5}$ | – | – |
| P | Dilute sol. | Fe | 1550 | $4.7 \times 10^{-5}$ | – | – |
| S | <0.64 | gr.satu.Fe | 1390–1560 | – | $2.8 \times 10^{-4}$ | 31 |
| S | ~1 | Fe | 1560–1670 | – | $4.9 \times 10^{-4}$ | 36 |
| Si | <2.5 | Fe | 1480 | $2.4 \times 10^{-5}$ | – | – |
| Si | <1.3 | Fe | 1540 | $3.8 \times 10^{-5}$ | – | – |
| Si | 1.5 | gr.satu.Fe | 1400–1600 | – | $2.4 \times 10^{-4}$ | 34 |

Units of κ: $J\ cm^{-1}s^{-1}K^{-1} \equiv 10^{-2}kg\ m\ s^{-3}K^{-1} \equiv 10^{-2}Wm^{-1}K^{-1}$

The effect of temperature on the thermal conductivities of iron, carbon steels and high alloy steels are shown in Fig. 4.40.

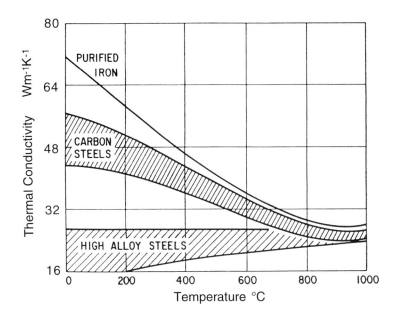

Fig. 4.40  Temperature dependence of thermal conductivity of purified iron and iron alloys.

## 4.12 THERMAL DIFFUSIVITY

Analogous to mass diffusivity, the thermal diffusivity $\alpha$ is defined as

$$\alpha = \frac{\kappa}{\rho C_p} \quad cm^2s^{-1} \tag{4.59}$$

where $\rho$ is density and $C_p$ molar heat capacity. In metals, the electrons migrate at much faster rates than the atoms; therefore the thermal diffusivity is much greater than the mass diffusivity.
For low-alloy steels at 1000°C:

$\alpha \approx 0.04$ $cm^2s^{-1}$
$D \approx 3 \times 10^{-12}$ $cm^2s^{-1}$ (substitutional)
$D \approx 3 \times 10^{-7}$ $cm^2s^{-1}$ (interstitial)

## REFERENCES

1. G.K. SIGWORTH and J.F. ELLIOTT, *Met. Sci.*, 1974, **8**, 298.
1a. R.P. SMITH, *J. Am. Chem. Soc.*, 1946, **68**, 1163.
2. J.F. ELLIOTT, M. GLEISER and V. RAMAKRISHNA, *Thermochemistry for steelmaking*, vol. II, Addison-Wesley Publishing Co., Reading, Massachusetts, 1963.
3. C. WELLS, *Trans. ASM*, 1938, **26**, 289.
4. J.C. SWARTZ, *Trans. AIME*, 1969, **245**, 1083.
5. F. VON NEUMANN and H. SCHENCK, *Arch. Eisenhüttenw*, 1959, **30**, 477.
6. E.T. TURKDOGAN and L.E. LEAKE, *J. Iron and Steel Inst.*, 1955, **179**, 39.
7. E.T. TURKDOGAN and R.A. HANCOCK, *ibid*, 1955, **179**, 155; 1956, **183**, 69.
8. J. CHIPMAN, R.M. ALFRED, L.W. GOTT, R.B. SMALL, D.M. WILSON, C.N. THOMSON, D.L. GUERSNEY and J.C. FULTON, *Trans. ASM*, 1952, **44**, 1215.
9. J. CHIPMAN and T.P. FLORIDIS, *Acta Met*, 1955, **3**, 456.
10. N.R. GRIFFING, W.D. FORGENG and G.W. HEALY, *Trans. AIME*, 1962, **224**, 148.
11. S.N. SINGH and K.E. BLAZEK, *J. Metals (AIME)*, 1974, **26**(10), 17.
12. M.M. WOLF and W. KURZ, *Metall. Transaction B*, 1981, **12B**, 85.
13. A.A. HOWE, *Applied Scientific Research*, 1987, **44**, 51.
14. H. YAMADA, T. SAKURAI and T. TAKENOUCHI, *Tetsu-to-Hagané*, 1990, **76**, 438.
15. M.M. WOLF, in *1st European Conference on Continuous Casting*, p. 2.489. Florence, Italy, Sept. 23–25, 1991.
16. L.S. DARKEN and R.W. GURRY, *J. Am. Chem. Soc.*, 1945, **67**, 1398; 1946, **68**, 798.
17. M.T. HEPWORTH, R.P. SMITH and E.T. TURKDOGAN, *Trans. Met. Soc. AIME*, 1966, **236**, 1278.
18. J.H. SWISHER and E.T. TURKDOGAN, *Trans. Met. Soc. AIME*, 1967, **239**, 426.
19. A. KUSANO, K. ITO and K. SANO, *Trans Iron and Steel Inst. Japan*, 1970, **10**, 78.
20. C.R. TAYLOR and J. CHIPMAN, *Trans. AIME*, 1943, 154, 228.
21. P.A. DISTIN, S.G. WHITEWAY and C.R. MASSON, *Can. Metall.Q.*, 1971, **10**, 13.
22. E.T. TURKDOGAN, *Ironmaking and Steelmaking*, 1993, **20**, 469.

23. T. Rosenqvist and B.L. Dunicz, *Trans. AIME*, 1952, **194**, 604; *J. Iron and Steel Inst.*, 1954, **176**, 37.
24. E.T. Turkdogan, S. Ignatowicz and J. Pearson, *J. Iron and Steel Inst.*, 1955, **180**, 349.
25. E.T. Turkdogan and G.J.W. Kor, *Metall. Transaction*, 1971, **2**, 1561.
26. D.C. Hilty and W. Crafts, *Trans. AIME*, 1952, **194**, 1307.
27. S. Bog and T. Rosenqvist, *Proc. Natl. Phys. Lab. Symp.*, 1958, No. 9, p. 6B.
28. A.E. Lord and N.A. Parlee, *Trans. Met. Soc. AIME*, 1960, **218**, 644.
29. D.L. Sponseller and R.A. Flinn, *Trans. Met. Soc. AIME*, 1964, **230**, 876.
30. P.K. Trojan and R.A. Flinn, *Trans. ASM*, 1961, **54**, 549.
31. P.J. Guichelaar, P.K. Trojan, T. Cluhan and R.A. Flinn, *Metall. Trans.*, 1971, **2**, 3305.
32. T.A. Engh, H. Midtgaard, J.C. Berke and T. Rosenqvist, *Scand. J. Metall.*, 1979, **8**, 195.
33. M. Speer and N. Parlee, *AFS Cast Metals Res. J.*, 1972, **8**, 122.
34. G.A. Irons and R.I.L. Guthrie, *Can. Metall. Quart.*, 1976, **15**, 325.
35. K.V. Gourishankar, M.K. Ranjbar and G.R. St.Pierre, *J. Phase Equilibria*, 1993, **14**, 601.
36. JANAF Thermochemical Tables – 1985 Supplement, *J. Phys. Chem. Reference Data*, 1985, **14**.
37. M. Nadif and C. Gatellier, *Rev. Metall.* CIT, 1986, **83**, p. 377.
38. S. Kobayashi, Y. Omori and K. Sanbongi, *Tetsu-to-Hagané*, 1970, **56**, 998.
39. K. Suzuki, A. Ejima and K. Sanbongi, *Tetsu-to-Hagané*, 1977, **63**, 585.
40. S. Gustaffsson, P.O. Mellberg and G. Ruist, *Thermodynamic behaviour of calcium in liquid iron*, Final Report ECSC, November 1982.
41. Q. Han, X. Zhang, D. Chen and P. Wang, *Metall. Trans. B.*, 1988, **19B**, 617.
42. E.B. Teplitskii and L.P. Vladimirov, *Russian Journal Phys. Chem.*, 1977, **51**, 490.
43. E.T. Turkdogan, *Iron & Steelmaker*, 1989, **16**(5), 61.
44. B.J. Keene, *A survey of extant data for the surface tension of iron and its binary alloys*, NPL Rep. DMA(A) 67, June 1983.
45. P. Kozakevitch, in *Surface phenomena of metals*, p. 223. Chem. Ind., London, 1968.
46. E.T. Turkdogan in *Foundry Processes – their chemistry and physics*, p. 53, Ed. S. Katz and C.F. Landefeld, Plenum Press, New York, 1988.
47. E. Selcuk and D.H. Kirkwood, *J. Iron and Steel Inst.*, 1973, **211**, 134.
48. L.D. Lucas, *Compt. Rend.*, 1959, **248**, 2336.
49. E.N. da C. Andrade, *Philo. Mag.*, 1934, **17**, 497, 698.
50. G. Cavalier, *Proc. Nat. Phys. Lab. No. 9*, 1958, vol. 2, 4D.
51. R.N. Barfield and J.A. Kitchener, J. Iron and Steel Inst., 1955, **180**, 324.

# CHAPTER 5

# Physicochemical Properties of Molten Slags

## 5.1 STRUCTURAL ASPECTS

Molten slags are ionic in nature consisting of positively charged ions known as **cations**, and negatively charged complex silicate, aluminate and phosphate ions known as **anions**.

The fundamental building unit in solid silica and molten silicates is the silicate tetrahedron $SiO_4^{-4}$. Each silicon atom is tetrahedrally surrounded by four oxygen atoms and each oxygen atom is bonded to two silicon atoms. The valency of silicon is +4 and that of oxygen is –2, therefore the silicate tetrahedron has 4 negative charges.

The three dimensional silicate structure repeats regularly in crystalline silica. In molten silica the $SiO_2$ tetrahedra are irregularly grouped, as illustrated in Fig. 5.1.

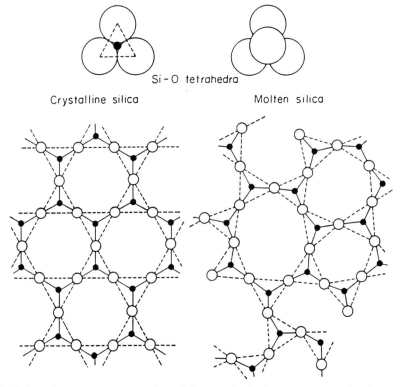

Fig. 5.1 Schematic representation of the tetrahedral arrangement of silica network in crystalline and molten silica.

138

The addition of metal oxides such as FeO, CaO, MgO, . . . to molten silica brings about a breakdown of the silicate network, represented in a general form by the reaction

$$> Si - O - Si < + MO \rightarrow 2 (> Si - O)^- + M^{2+} \qquad (5.1)$$

The cations are dispersed within the broken silicate network. In MO–$SiO_2$ melts the atom ratio O/Si > 2, therefore part of the oxygen atoms are bonded between two silicon atoms and part to only one silicon atom. Partial depolymerisation of the silicate network with the addition of a metal oxide MO is illustrated in Fig. 5.2.

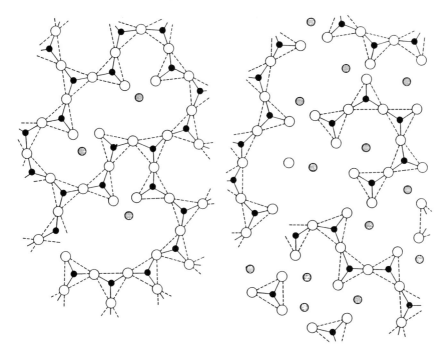

Fig. 5.2   Schematic representation of depolymerisation of the silicate network with the dissolution of metal oxides in silicate melts.

In highly basic slags with molar ratio MO/$SiO_2$ > 2, the silicate network completely breaks down to individual $SiO_4$ tetrahedra intermixed with cations $M^{2+}$ and some oxygen ions $O^{2-}$.

At low concentrations, $Al_2O_3$ behaves like a network-modifying oxide and forms aluminium cations $Al^{3+}$. At high concentrations, the aluminium enters the tetrahedral structure isomorphous with silicon.

This process may be schematically represented by the reaction

$$> Si - O - Si < + MAlO_2 \rightarrow > Si - O - \overset{\mid}{Al} - O -Si < + M^+ \qquad (5.2)$$
$$\underset{O}{\mid}$$

The cation $M^+$ is located in the vicinity of Al–O bonding to preserve the local charge balance. At low concentrations of phosphorus in steelmaking slags, the phosphate ions $PO_4^{3-}$ are incorporated in the silicate network. In steelmaking slags, the sulphur exists as a sulphide ion $S^{2-}$. The sulphate ions $SO_4^{2-}$ exist in slags only under highly oxidising conditions and in the absence of iron or any other oxidisable metal.

Although molten slags are ionised, the slag composition can be represented in terms of the constituent oxides, e.g. $CaO$, $FeO$, $SiO_2$, $P_2O_5$ . . . The thermodynamic activity of an ion in the slag cannot be determined. However, the activity of an oxide dissolved in molten slag, forming $M^{2+}$ and $O^{2-}$ ions, can be determined experimentally and the following equality can be written

$$MO \rightarrow M^{2+} + O^{2-} \tag{5.3}$$

$$\frac{a_{MO}}{(a_{MO})^\circ} = \frac{(a_{M2+} \times a_{O2-})}{(a_{M2+} \times a_{O2-})^\circ} \tag{5.4}$$

where the superscript $^\circ$ refers to the standard state which is usually pure solid or liquid oxide.

## 5.2 SLAG BASICITY

For steelmaking slags of low phosphorus content, the slag basicity has traditionally been represented by the mass concentration ratio

$$V = \frac{\%CaO}{\%SiO_2} \tag{5.5}$$

For slags containing high concentrations of $MgO$ and $P_2O_5$, as in some laboratory experiments, the basicity may be defined by the following mass concentration ratio, with the assumption that on a molar basis the concentrations of $CaO$ and $MgO$ are equivalent. Similarly, on a molar basis $\frac{1}{2}$ $P_2O_5$ is equivalent to $SiO_2$.

$$B = \frac{\%CaO + 1.4\,\%MgO}{\%SiO_2 + 0.84\%\ P_2O_5} \tag{5.6}$$

For slags containing $MgO < 8\%$ and $P_2O_5 < 5\%$, the basicity $B$ is essentially directly proportional to $V$.

$$B = 1.17V \tag{5.7}$$

Another measure of slag basicity is the difference between the sum of the concentrations of basic oxides and acidic oxides.

$$(\%CaO + \%MgO + \%MnO) - (\%SiO_2 + \%P_2O_5 + \%TiO_2) \tag{5.8}$$

This formulation of slag basicity is not used very often.

For the calcium aluminate type of slags used in steel refining in the ladle furnace, the slag basicity used in some German and Japanese publications, is defined by the ratio

$$\frac{\% \text{CaO}}{\% \text{SiO}_2 \times \% \text{Al}_2\text{O}_3} \tag{5.9}$$

However, such a ratio becomes meaningless at low concentrations of either $\text{SiO}_2$ or $\text{Al}_2\text{O}_3$. For the ladle furnace slag the basicity may be defined by the following mass concentration ratio

$$B_{LF} = \frac{\% \text{CaO} + 1.4 \times \% \text{MgO}}{\% \text{SiO}_2 + 0.6 \times \% \text{Al}_2\text{O}_3} \tag{5.10}$$

The slag basicities as defined above are for the compositions of molten slags. In practice, the steelmaking slags often contain undissolved CaO and MgO. The chemical analyses of such slag samples without correction for undissolved CaO and MgO, will give unrealistic basicities which are much higher than those in the molten part of the slag.

There has been a trend in recent years to relate some physicochemical properties of slags, such as sulphide capacity, phosphate capacity, carbide capacity etc., to the so-called optical slag basicity. Such attempts to generalise the physicochemical properties of silcates, phosphates, aluminates etc., into a single rationale invariably lowers the accuracy of representing the composition dependence of the equilibrium constants of gas–slag–metal reactions. As is seen from the plot in Fig. 5.3, there is no meaningful

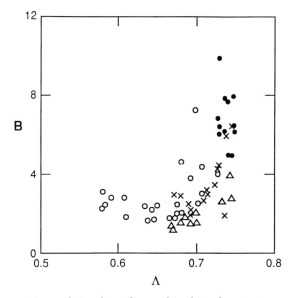

Fig. 5.3  Compositions of simple and complex slags do not give a meaningful correlation between optical basicity $\Lambda$ and concentration-based slag basicity $B$.

correlation between the so-called optical basicity $\Lambda$ and the concentration-based slag basicity $B$ for simple and complex slags, to which references will be given later in this chapter.

## 5.3 SELECTED PHASE DIAGRAMS OF BINARY AND TERNARY OXIDE SYSTEMS

Pertinent to the steelmaking type of slags, a few phase diagrams of binary and ternary oxide systems are given here from the data compiled by Levin et al.[1]

Iron oxide dissolves in slags in two valency states: divalent iron cations $Fe^{2+}$ and trivalent iron cations $Fe^{3+}$. The ratio $Fe^{3+}/Fe^{2+}$ depends on temperature, oxygen potential and slag composition; this is discussed later in this chapter. In the formulation of the equilibrium constants of slag–metal reactions and the thermodynamic activities of oxides in slags, the total iron dissolved in the slag as oxides is usually converted to the stoichiometric formula FeO and denoted by $Fe_tO$, thus

$$\%Fe_tO = \%FeO \text{ (analysed)} + 0.9 \times \%Fe_2O_3 \text{ (analysed)} \qquad (5.11a)$$

$$\text{or} \qquad \%Fe_tO = 1.286 \times \%Fe \text{ (total as oxides)} \qquad (5.11b)$$

For the sake of simplicity in printing, the subscript $t$ will be omitted in all the subsequent equations and diagrams.

### 5.3.1 BINARY OXIDE SYSTEMS

The simplest oxide system is that of FeO–MnO where there is a complete series of liquid and solid solutions (Fig. 5.4).

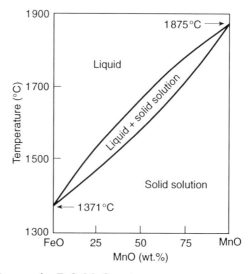

Fig. 5.4   Phase diagram for FeO–MnO system.

In the CaO–MgO system (Fig. 5.5) there is a small solid solubility at the terminal regions at the eutectic temperatures ~ 2300°C. Dolomite, $MgCa(CO_3)_2$, is an important raw material for use in steelmaking furnaces as a basic refractory material. The dolomite is calcined at temperatures in the order of 1700°C to yield a material less subject to deterioration by moisture during storage.

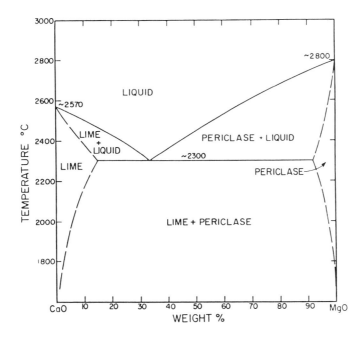

Fig. 5.5   Phase diagram for CaO–MgO system.

In considering the silicate binary systems, mention should be made of the polymorphic forms of silica. The following are the major phase trans-formations of silica: melting temperature 1723°C; cristobalite → tridymite transformation 1470°C; tridymite → quartz transformation 867°C. However, tridymite is not a true polymorphic form of pure silica. It has been demonstrated[2–4] that tridymite is formed within the temperature range 867–1470°C only when some metal ions dissolve in silica. It is for this reason that in all the metal oxide–silica systems tridymite appears as a stable phase over the indicated temperature range. In most of the silicate systems, except those with alkali oxides and alumina, there is a liquid miscibility gap as seen from the phase diagrams in Figs. 5.6, 5.7 and 5.8. Another characteristic feature is that there is a negligible amount of solid-solution formation in most of the silicate systems.

In the CaO–SiO$_2$ system (Fig. 5.6) there are two phases with congruent melting points: $\alpha$-2CaO·SiO$_2$ and CaO·SiO$_2$; two with incongruent melt-ing points: 3CaO·2SiO$_2$ and 3CaO·SiO$_2$, the latter being stable only

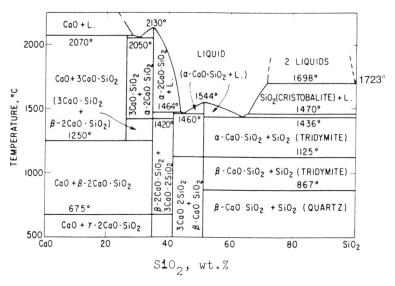

Fig. 5.6   Phase diagram for CaO–SiO$_2$ system.

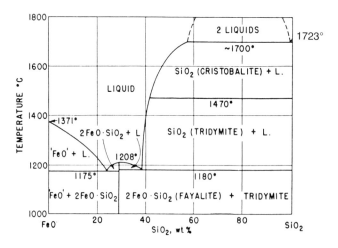

Fig. 5.7   Phase diagram for FeO–SiO$_2$ system.

within the temperature range 1250–2070°C. One characteristic feature of the di-calcium silicate is that on conversion from β to γ crystalline modification at 675°C, there is about a 12 percent volume expansion causing the solid mass to become powdery. In electric furnace practice, the reducing slag used has a composition similar to that of di-calcium silicate, and because of the transformation from β to γ form, on cooling the slag becomes powdery; in industry this type of slag is often referred to as the 'white falling slag'. The formation of γ-2CaO·SiO$_2$ in blast furnace slags is highly undesirable if the slag is to be used as an aggregate for constructional purposes.

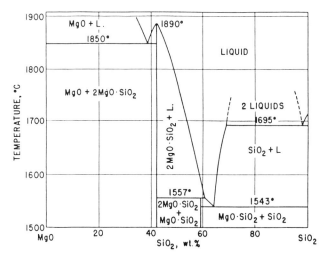

Fig. 5.8   Phase diagram for MgO–SiO$_2$ system.

The phase diagram for the FeO–SiO$_2$ system in Fig. 5.7 is that in equilibrium with solid or liquid iron; this is not a true binary system. The amount of ferric oxide in the melt along the liquidus curve, in equilibrium with iron, decreases from about 12 percent at the FeO corner to about 1 percent at silica saturation.

The phase diagram for the MgO-SiO$_2$ system is given in Fig. 5.8. Below 40% SiO$_2$ the eutectic melting occurs at 1850°C indicating that magnesia, even in the presence of appreciable amounts of silica, is a good refractory material. The refractory oxides and some silicates are often called by their mineralogical names: MgO (magnesia)-periclase; 2 MgO·SiO$_2$ – forsterite; MgO·SiO$_2$ – protoenstatite.

The phase diagram for the CaO–Al$_2$O$_3$ system is given in Fig. 5.9. The prefused or sintered calcium aluminates made for flux additions to the steel ladle usually contain 42 to 48% Al$_2$O$_3$, which is within the region of low liquidus temperatures.

5.3.2 TERNARY AND QUATERNARY OXIDE SYSTEMS

Most steelmaking slags consist primarily of CaO, MgO, SiO$_2$ and FeO. In low-phosphorus steelmaking practices, the total concentration of these oxides in liquid slags is in the range 88 to 92%. Therefore, the simplest type of steelmaking slag to be considered is the quaternary system CaO–MgO–SiO$_2$–FeO.

First, let us consider the ternary system CaO–SiO$_2$–FeO; the liquidus isotherms of this system is shown in Fig. 5.10. The isothermal section of the composition diagram in Fig. 5.11 shows the phase equilibria at 1600°C.

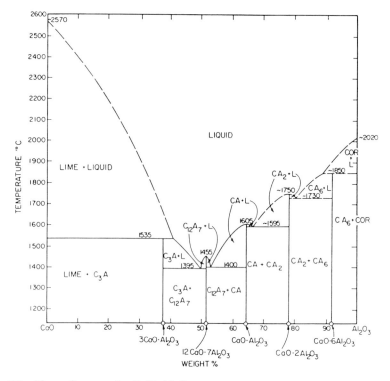

Fig. 5.9   Phase diagram for CaO–Al$_2$O$_3$ system.

There are four two-phase regions where, as depicted by dotted lines, the melt is saturated with SiO$_2$, 2CaO·SiO$_2$, 3CaO·SiO$_2$ or CaO; two three-phase regions (2CaO·SiO$_2$ + 3 CaO·SiO$_2$ + liquid) and (3CaO·SiO$_2$ + CaO + liquid); and one liquid phase region.

Magnesia is another important ingredient of steelmaking slags, which are invariably saturated with MgO to minimise slag attack on the magnesia refractory lining of the furnace.

The effect of MgO on the solubility of calcium silicates and calcium oxide is shown in Fig. 5.12 for the system (CaO + MgO)–SiO$_2$–FeO, in equilibrium with liquid iron at 1600°C. The broken-line curve delineates the region of saturation of molten slag with solid calcium (magnesium) silicates and solid magnesio-wustite (MgO–FeO solid solution). Effects of the concentrations of MgO and FeO on the solubility of CaO in 2CaO·SiO$_2$-saturated slags are shown in Fig. 5.13.

Tromel *et al.*[5] have made a detailed study of the solubility of MgO in iron–calcium silicate melts. The solubility data are given in Fig. 5.14 as iso-MgO concentrations for CaO–MgO–SiO$_2$–FeO slags saturated with olivine (Mg, Fe)$_2$SiO$_4$, and magnesio-wustite in equilibrium with liquid iron at 1600°C. Point *A* is for the ternary system Mg–Fe–O and *B* for Mg–Ca–O. Along the composition path *AB*, molten slag is saturated with both olivine

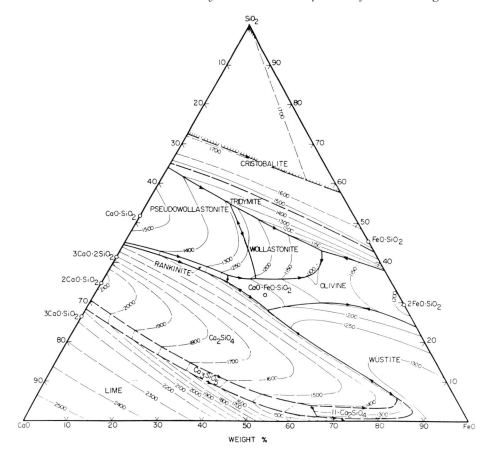

Fig. 5.10   Liquidus isotherms of CaO–SiO$_2$–FeO system.

and Mg(Fe)O. Along the composition path *CD*, the slag is saturated with both pyroxene (Ca,Mg) silicates and magnesio-wustite.

Below the dotted curve *BACD* in Fig. 5.15 for double saturations, the curves for 10 to 60% FeO are the MgO solubilities in the slag as magnesio-wustite. The solubility data obtained in other experimental work[6–8] are in general accord with the data of Tromel *et al.* in Fig. 5.15. In the work of Leonard and Herron,[7] it was also found that the addition of CaF$_2$ up to about 10% had no perceptible effect on the solubility of MgO in steelmaking type slags.

The shift in the solubility isotherms at 1600°C with the addition of 9 and 15% MnO to the system CaO–SiO$_2$–FeO, co-existing with liquid iron, is shown in Fig. 5.16 as determined by Gorl *et al.*[9]

Pertinent to the compositions of neutral ladle slags, the phase equilibria in part of the system CaO–Al$_2$O$_3$–SiO$_2$ at 1600°C is shown in Fig. 5.17.

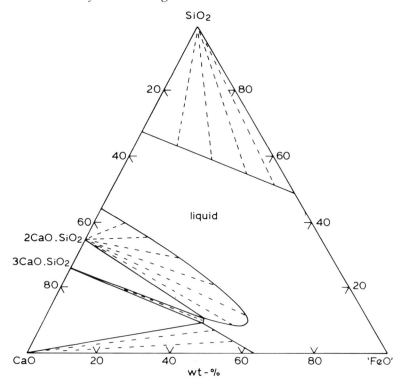

Fig. 5.11   Phase equilibria in the system CaO–SiO$_2$–FeO in equilibrium with liquid iron at 1600°C.

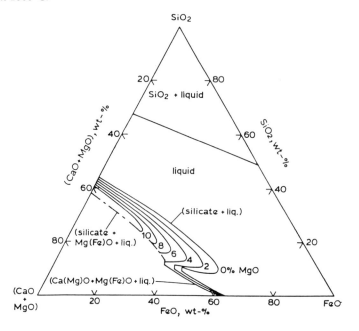

Fig. 5.12   Effect of MgO (wt.%) on the solubility isotherms at 1600°C in the system (CaO + MgO)–SiO$_2$–FeO in equilibrium with liquid iron.

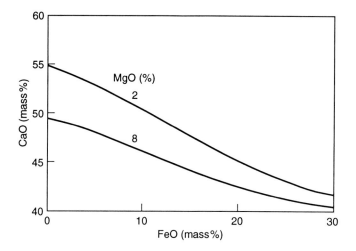

Fig. 5.13 Solubility of CaO in (CaO–MgO–SiO$_2$–FeO) slags saturated with 2CaO·SiO$_2$ at 1600°C.

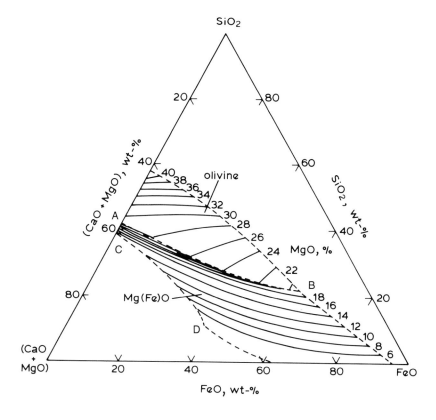

Fig. 5.14 Iso–MgO concentrations (wt.%) for CaO–MgO–SiO$_2$–FeO slags saturated with olivine and magnesio-wustite in equilibrium with liquid iron at 1600°C.

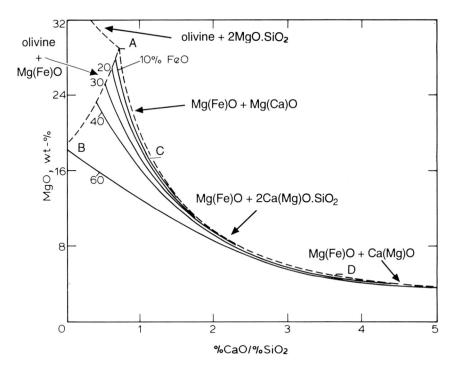

Fig. 5.15   Solubility of MgO, as magnesio-wustite, in the system CaO–MgO–SiO₂–FeO at 1600°C as a function of slag basicity and FeO concentration, derived from the data in Fig. 5.14.

## 5.4 THERMODYNAMIC ACTIVITIES OF OXIDES IN SIMPLE AND COMPLEX SLAGS

Thermodynamic activities of oxides dissolved in molten slags are relative to pure liquid or solid oxides as the standard state; that is, for pure oxide the activity $a_{MO} = 1$.

Reference may be made to a previous publication by the author,[10] for compiled experimental data on the thermodynamic activities in binary and ternary oxide systems. In this chapter the activity data are given for a few ternary and multicomponent systems which are closely related to steelmaking slags.

### 5.4.1 ACTIVITIES IN CaO–FeO–SiO₂ SYSTEM

The oxide activities in the CaO–FeO–SiO₂ melts in equilibrium with liquid iron at about 1550°C are given in Fig. 5.18. The iso-activity curves in the top diagram represent the experimentally determined activities of iron oxide with respect to liquid FeO.[11,12] The activities of CaO and SiO₂ with

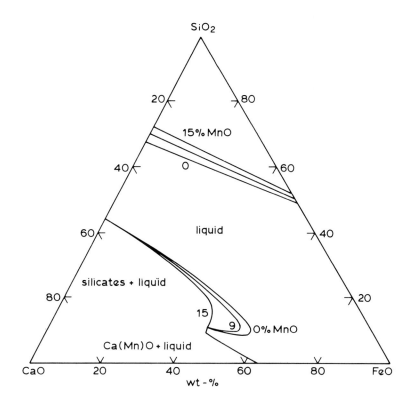

Fig. 5.16   Shift in the solubility isotherms at 1600°C with the addition of 9 and 15% MnO to the system CaO–SiO$_2$–FeO; the total CaO, SiO$_2$ and FeO contents are related to 100%.

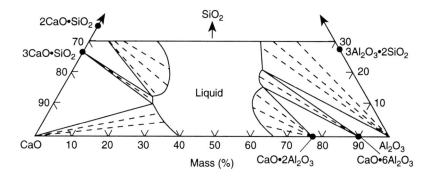

Fig. 5.17   Phase equilibria in the system CaO–Al$_2$O$_3$–SiO$_2$ at 1600°C.

respect to solid oxides were calculated from the FeO activities by Gibbs–Duhem integration.[12]

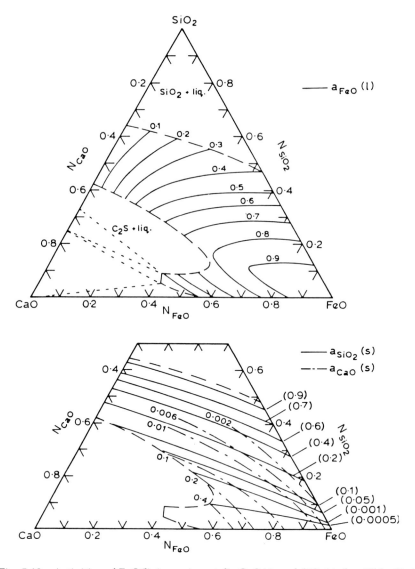

Fig. 5.18    Activities of FeO(l) (experimental), CaO(s) and SiO₂(s) (by Gibbs-Duhem integration) in CaO–FeO–SiO₂ melts in equilibrium with liquid iron at 1550°C. From Refs. 11, 12.

### 5.4.2 ACTIVITIES IN CaO–Al₂O₃–SiO₂ SYSTEM

Rein and Chipman[13] measured the activity of silica in the CaO–Al₂O₃–SiO₂ system; from these experimental data they calculated the activities of CaO and Al₂O₃ by Gibbs–Duhem integration. The salient features of these oxide activities at 1600°C, with respect to solid oxides, are shown in Fig. 5.19 for the mass ratios of CaO/Al₂O₃ = 2/3 and 3/2; the compositions of ladle slags are well within the range given in Fig. 5.19.

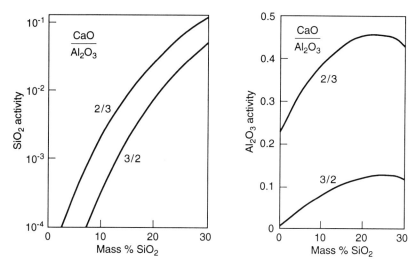

Fig. 5.19   Silica and alumina activities, with respect to solid oxides, in CaO–Al$_2$O$_3$–SiO$_2$ melts at 1600°C, derived from experimental data of Rein and Chipman.[13]

The ratio of the activities $(a_{Al_2O_3})^{1/3}/(a_{CaO})$, within the entire liquid composition range up to 30% SiO$_2$, is shown in Fig. 5.20.

### 5.4.3 ACTIVITIES IN MnO–Al$_2$O$_3$–SiO$_2$ SYSTEM

In the deoxidation of steel with the ladle addition of silicomanganese and aluminium together, the deoxidation product is molten manganese aluminosilicate with the mass ratio MnO/SiO$_2$ at about 1:1 and containing 10 to 45% Al$_2$O$_3$. The activities of oxides in the MnO–Al$_2$O$_3$–SiO$_2$ melts at 1550° and 1650°C were computed by Fujisawa and Sakao[14] from the available thermochemical data on the system. They also determined experimentally the activities of MnO and SiO$_2$ by selected slag-metal equilibrium measurements and found a close agreement with the computed data.

The activities of Al$_2$O$_3$ and SiO$_2$, with respect to solid oxides, are plotted in Fig. 5.21 for melts with mass ratio of MnO/SiO$_2$ = 1. For melts containing up to 30% Al$_2$O$_3$ the activity of MnO remains essentially unchanged at about 0.1 then decreases to about 0.05 at 40% Al$_2$O$_3$.

### 5.4.4 ACTIVITY COEFFICIENT OF FeO IN SLAGS

With hypothetical pure liquid FeO as the standard state, the activity of iron oxide is derived from the concentration of dissolved oxygen in liquid iron that is in equilibrium with the slag. For the reaction equilibrium

$$FeO(l) = Fe + [O] \qquad (5.12)$$

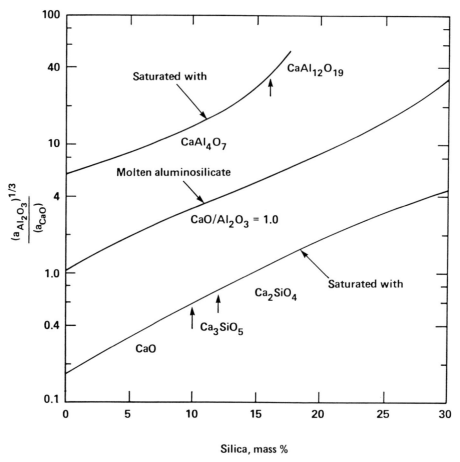

Fig. 5.20   Effect of slag composition on the activity ratio $(a_{Al_2O_3})^{1/3}/a_{CaO}$ for the system CaO–Al$_2$O$_3$–SiO$_2$ at 1600°C.

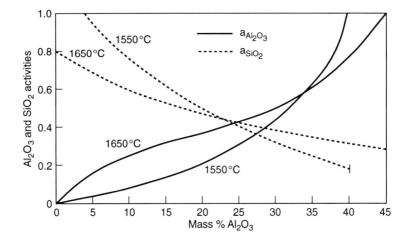

Fig. 5.21   Al$_2$O$_3$ and SiO$_2$ activities in MnO–Al$_2$O$_3$–SiO$_2$ system for mass ratio of MnO/SiO$_2$ = 1, derived from data compiled by Fujisawa & Sakao.[14]

$$K_O = \frac{[a_O]}{a_{FeO}} \qquad (5.12a)$$

where $a_O = [\%O]f_O$, using $\log f_O = -0.1 \times [\%O]$. The temperature dependence of the equilibrium constant $K_O$ given below is that derived from the free energy equation (4.36a).

$$\log K_O = - \frac{5730}{T} + 2.397 \qquad (5.12b)$$

In the past, we have generated a wide variety of formulations to represent the composition dependence of the iron oxide activity or activity coefficient in complex slags. In the present reassessment of this property of the slag the author came to the conclusion that, within the limits of uncertainty of the experimental data on slag-metal reaction equilibrium, there is a decisive correlation between the activity coefficient of FeO and the slag basicity B as shown in Fig. 5.22; the experimental data used are from numerous independent studies.[11,15-20] It is evident that $\gamma_{FeO}$ reaches a peak at a basicity of about $B = 1.8$. It should be pointed out once again that the concentration of iron oxide is for total iron as oxides in the slag represented by the stoichiometric formula FeO as in equation (5.11b).

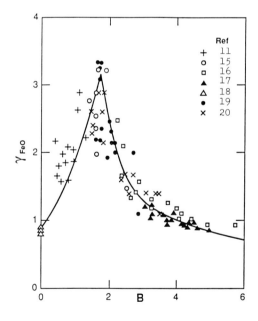

Fig. 5.22   Effect of slag basicity on the activity coefficient of iron oxide in simple and complex slags at 1600°C.

The experimental data cited in Fig. 5.22 are for simple and complex slags which differ considerably in their compositions. Some data are for silicate melts with no $P_2O_5$ and basicities of $B = 0$ to 3. On the other extreme the equilibrium data used are for slags saturated with CaO and contain $P_2O_5$ in the range 1 to 20%. It should be noted that the correlation in Fig. 5.22 is for slags containing CaO, $SiO_2$, FeO<40%, MnO<15%, MgO<20% and $P_2O_5$<20%.

There are several other experimental data for temperatures of 1530° to 1750°C which give $\gamma_{FeO}$ values that are in general accord with the relationship shown in Fig. 5.22 for 1600°C.

### 5.4.5 ACTIVITY COEFFICIENT OF MnO IN SLAGS

The slag/metal manganese distribution ratio is governed by two interrelated reaction equilibria. One reaction is oxidation of Mn with FeO in the slag.

$$(FeO) + [Mn] = (MnO) + Fe \tag{5.13}$$

$$K_{FeMn} = \frac{a_{MnO}}{a_{FeO}[\%Mn]} \tag{5.13a}$$

where the oxide activities are with respect to pure liquid oxides.

The second reaction to be considered is

$$[Mn] + [O] = (MnO) \tag{5.14}$$

$$K_{Mn} = \frac{a_{MnO}}{[\%Mn][a_O]} \tag{5.14a}$$

From a critical reassessment of the experimental data of numerous studies, the author[21] derived the following equations for the temperature dependence of the equilibrium constants $K_{FeMn}$ and $K_{Mn}$.

$$\log K_{FeMn} = \frac{7452}{T} - 3.478 \tag{5.13b}$$

$$\log K_{Mn} = \frac{13,182}{T} - 5.875 \tag{5.14b}$$

The ratio of the activity coefficients $\gamma_{FeO}/\gamma_{MnO}$ in slags is derived from the equilibrium data cited in Fig. 5.22 for simple and complex slags and the equilibrium constant $K_{FeMn}$ thus

$$\frac{\gamma_{FeO}}{\gamma_{MnO}} = \frac{1}{K_{FeMn}} \left\{ \frac{N_{MnO}}{N_{FeO}[\%Mn]} \right\} \tag{5.15}$$

where Ns are mol fractions of the oxides. The results are given in Fig. 5.23. There is a sharp decrease in the ratio of the activity coefficients as the basicity $B$ increases from 1.5 to 2.0. At basicities above 2.5, the ratio $\gamma_{FeO}/\gamma_{MnO}$ is essentially constant at about 0.63.

The activity coefficient of MnO, with respect to pure liquid MnO as the standard state, is derived from the data cited above using the relation

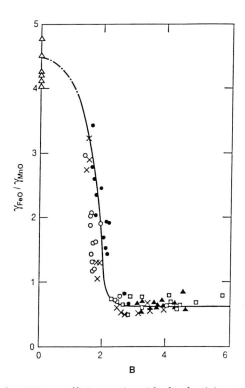

Fig. 5.23   Variation of activity coefficient ratio with slag basicity at 1600°C.

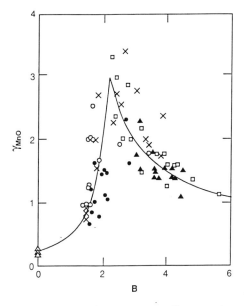

Fig. 5.24   Effect of slag basicity on the activity coefficient of manganese oxide in simple and complex slags at 1600°C.

$$\gamma_{MnO} = K_{Mn} \left\{ \frac{[\%Mn][a_O]}{N_{MnO}} \right\} \tag{5.16}$$

The variation of $\gamma_{MnO}$ with slag basicity shown in Fig. 5.24 is similar to that of $\gamma_{FeO}$. It should be noted that to maintain consistency in the interpretation of the data, the curve drawn in Fig. 5.24 is that derived from the combination of curves in Figs. 5.22 and 5.23.

## 5.5 Oxygen Potential Diagram

The reducibility of a metal oxide relative to other oxides, or the oxidisability of a metal relative to other metals, can readily be assessed from the free energy data.

For the oxidation reaction involving pure metals and metal oxides in their standard states, i.e. $a_M = 1$, $a_{MO} = 1$

$$2M + O_2 = 2MO \tag{5.17}$$

the isothermal equilibrium constant is

$$K = \frac{1}{p_{O_2}} \tag{5.18}$$

where $p_{O_2}$ is the equilibrium oxygen partial pressure for which the standard state is 1 atm at the temperature under consideration. The standard free energy change is

$$\Delta G^\circ = - RT \ln K = RT \ln p_{O_2} \tag{5.19}$$

which is also called the **oxygen potential**.

In an earlier study, Richardson and Jeffes[22] compiled the free energy data on metal oxides which were then available, and presented the data as an oxygen potential diagram (Fig. 5.25) that is reproduced from their paper.

The oxides for which the oxygen potential lines are above that of CO can be reduced by carbon. As the affinity of the metals for oxygen increases, i.e. $\Delta G^\circ$ decreases, the temperature of reduction of the oxides by carbon increases.

For easy conversion of oxygen potentials to the corresponding values of $p_{O_2}$ or to the equilibrium ratios of $H_2/H_2O$ and $CO/CO_2$, appropriate scales are included in the enclosed oxygen potential diagram for various oxides of metallurgical interest.

Scale for $p_{O_2}$:

Lines drawn from the point $O$ on the ordinate for the absolute zero temperature through the points marked on the right hand side of the diagram give the isobars. For example, for the Fe–FeO equilibrium the

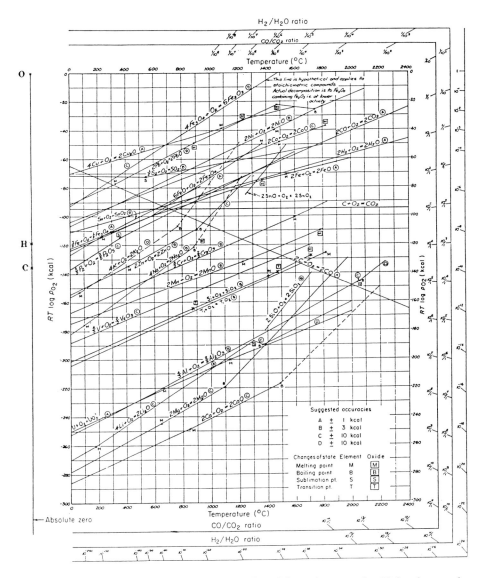

Fig. 5.25   Oxygen potential diagram, reproduced from the paper by Richardson and Jeffes.[22]

oxygen potential is −82 kcal (−343 kJ) at 1200°C. By drawing a line passing through this point and the point *O*, the oxygen partial pressure of about 7 × 10⁻¹³ atm is read off the $p_{O_2}$ scale.

Scale for CO/CO₂:

Draw the line from point *C* to a point on the oxygen potential; the extension of this line gives the corresponding equilibrium CO/CO₂ ratio.

Scale for H₂/H₂O:

Same as above by using point *H*.

## 5.6 Gas Solubilities in Slags

### 5.6.1 solubility of $H_2O$

In acidic melts, $H_2O$ vapour reacts with double bonded oxygen and de-polymerises the melt, thus

$$(\Rightarrow Si - O - Si \Leftarrow) + H_2O = 2\ (\Rightarrow Si - OH)$$

In basic melts, $H_2O$ reacts with free oxygen ions

$$(O^{2-}) + H_2O = 2(OH)^-  \tag{5.21}$$

Both for acidic and basic melts the overall reaction is represented by

$$(O^*) + H_2O = 2(OH^*)  \tag{5.22}$$

where O* represents double or single bonded oxygen, or $O^{2-}$, and OH* is single bonded to silicon or as a free ion. The equilibrium constant for a given melt composition is

$$C_{OH} = \frac{(ppm\ H_2O)}{(p_{H2O})^{1/2}}  \tag{5.23}$$

where $p_{H2O}$ is the vapour partial pressure.

The solubilities of $H_2O$ (in units of mass ppm $H_2O$) at 1 atm pressure of $H_2O$ vapour in $CaO–FeO–SiO_2$ and $CaO–Al_2O_3–SiO_2$ at 1550°C, measured by Iguchi et al.[23] are given in Fig. 5.26.

### 5.6.2 solubility of $N_2$

Nitrogen dissolves in molten slags as a nitride ion $N^{3-}$ only under reducing conditions

$$\tfrac{1}{2}N_2(g) + \tfrac{3}{2}(O^{2-}) = (N^{3-}) + \tfrac{3}{4}O_2(g)  \tag{5.24}$$

for which the equilibrium constant (known as nitride capacity) is

$$C_N = (\%N)\ \frac{p_{O2}^{3/4}}{p_{N2}^{1/2}}  \tag{5.25}$$

Many studies have been made of the solubility of nitrogen in $CaO–Al_2O_3$ and $CaO–Al_2O_3–SiO_2$ melts in the 1970s. These were reviewed by the author in a previous publication.[10] Reference should be made also to a subsequent work done by Ito and Fruehan[24] on the nitrogen solubility in the $CaO–Al_2O_3–SiO_2$ melts. They showed that the nitride capacity of the aluminosilicate melts increases with a decreasing activity of CaO as shown in Fig. 5.27.

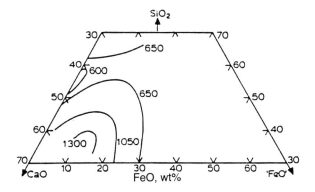

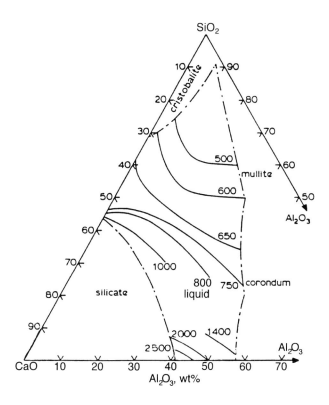

Fig. 5.26 Hydroxyl capacity, $C_{OH}$ ppm $H_2O$ atm$^{-1/2}$ of CaO–FeO–SiO$_2$ and CaO–Al$_2$O$_3$–SiO$_2$ melts at 1550°C. From Ref. 23.

*5.6.2a Slag/metal distribution ratio (N)/[N]*

For Al-killed steels the corresponding equilibrium partial pressure of oxygen is

$$p_{O_2} \text{ (atm)} = \frac{1.22 \times 10^{-16}}{[\% \text{Al}]^{4/3}} \quad \text{at } 1600°C$$

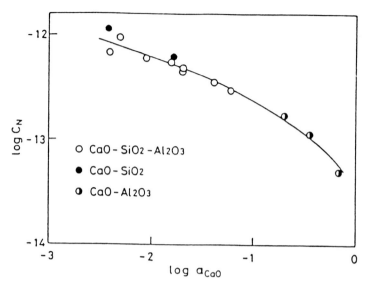

Fig. 5.27   The relation between the nitride capacity and the activity of lime for CaO–SiO$_2$–Al$_2$O$_3$ slag at 1550°C. From Ref. 24.

From the solubility data for nitrogen we have

$$p_{N_2}^{1/2}(atm^{1/2}) = 21.9[\%N] \text{ at } 1600°C$$

Substituting these in the equation for $C_N$ gives

$$C_N = \frac{(\%N)}{[\%N][\%Al]} \times 5.3 \times 10^{-14}$$

In lime-rich aluminate ladle slags of low SiO$_2$ content, the lime activity is $a_{CaO} = 0.5$ for which $C_N = 10^{-13}$. These give the following equilibrium relation

$$\frac{(\%N)}{[\%N]} = 1.9 \times [\%Al]$$

| Dissolved [%Al] | (%N)/[%N] |
|---|---|
| 0.005 | 0.0095 |
| 0.02 | 0.038 |
| 0.06 | 0.114 |
| 0.10 | 0.190 |

The equilibrium nitrogen distribution ratio between slag and steel is very low even at high aluminium contents, well above the practical range. The slag/steel mass ratio in the ladle is also low, about 1/100. For these reasons, liquid steel in the ladle cannot be de-nitrogenised by the slag that is usable in industry in steel refining.

5.6.3 SOLUBILITY OF $S_2$

Sulphur-bearing gases dissolve in molten slags as sulphide ions ($S^{2-}$) under reducing conditions, and as sulphate ions ($SO_4^{2-}$) under highly oxidising conditions. In steelmaking the oxygen potential is not high enough for the solution of sulphur as sulphate ions, therefore we need to consider only the sulphide reaction.

Whether the sulphur-bearing species is primarily $H_2S$ or $SO_2$, there is a corresponding equilibrium value of $p_{S_2}$ depending on the temperature and gas composition. It is convenient to consider the reaction in a general form as

$$\tfrac{1}{2}S_2(g) + (O^{2-}) = (S^{2-}) + \tfrac{1}{2}O_2(g) \tag{5.26}$$

For a given slag composition the equilibrium relation is represented by

$$C_S = (\%S) \left( \frac{p_{O_2}}{p_{S_2}} \right)^{1/2} \tag{5.27}$$

where $p$s are equilibrium gas partial pressures. The equilibrium constant $C_S$ is known as the sulphide capacity of the slag. The value of $C_S$ depends on slag composition and temperature.

Experimentally determined sulphide capacities of binary oxide melts are shown in Fig. 5.28. References to experimental data are given in Ref. 10.

For the slag–metal system, the sulphur reaction is formulated in terms of the activities ($\cong$ concentrations) of sulphur and oxygen dissolved in the steel.

$$[S] + (O^{2-}) = (S^{2-}) + [O] \tag{5.28}$$

For low-alloy steels

$$k_S = \frac{(\%S)}{[\%S]} [\%O] \tag{5.28a}$$

where the equilibrium constant $k_S$ depends on slag composition and temperature.

Conversion of $p_{O_2}/p_{S_2}$ to $[\%O]/[\%S]$:

The free energies of solution of $O_2$ and $S_2$ in liquid low alloy steel are given below; see Table 4.3 in Chapter 4.

$$\tfrac{1}{2}S_2 = [S] \qquad \Delta G_S = -135{,}060 + 23.43T \text{ J}$$
$$\tfrac{1}{2}O_2 = [O] \qquad \Delta G_O = -115{,}750 - 4.63T \text{ J}$$

For the reaction equilibrium

$$\tfrac{1}{2}S_2 + [O] = \tfrac{1}{2}O_2 + [S]$$

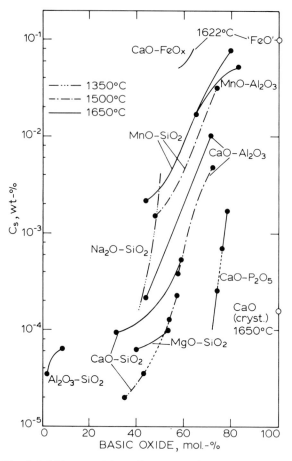

Fig. 5.28　Sulphide capacities of binary oxide melts. From Ref. 10.

the standard free energy change is

$$\Delta G° = \Delta G_S - \Delta G_O = -19{,}310 + 28.06T \text{ J} \tag{5.29}$$

$$K = \left(\frac{p_{O_2}}{p_{S_2}}\right)^{1/2} \frac{[\%S]}{[\%O]} = \frac{1009}{T} - 1.466 \tag{5.29a}$$

For steelmaking temperatures an average value of $K$ is 0.133; with this conversion factor the following is obtained.

$$\left(\frac{p_{O_2}}{p_{S_2}}\right)^{1/2} = 0.133 \frac{[\%O]}{[\%S]}$$

With this substitution the values of $C_S$ are converted to $k_S$.

$$7.5 \times C_S = k_S = \frac{(\%S)}{[\%S]} \, [\%O] \tag{5.30}$$

A slag of high $k_S$ value and steel deoxidation to low levels of [O] are

necessary conditions for steel desulphurisation. This subject is discussed in more detail later.

### 5.6.4 SOLUBILITY OF $O_2$

Oxygen dissolves in molten slags by oxidising the divalent iron ions to the trivalent state.
Gas-slag reaction:

$$\tfrac{1}{2}O_2(g) + 2(Fe^{2+}) = 2(Fe^{3+}) + (O^{2-}) \tag{5.31}$$

Slag-metal reaction:

$$Fe° + 4(Fe^{3+}) + (O^{2-}) = 5(Fe^{2+}) + [O] \tag{5.32}$$

These reactions provide the mechanism for oxygen transfer from gas to metal through the overlaying slag layer. Partly for this reason, the steel reoxidation will be minimised by maintaining a low concentration of iron oxide in the ladle slag, tundish and mould fluxes.

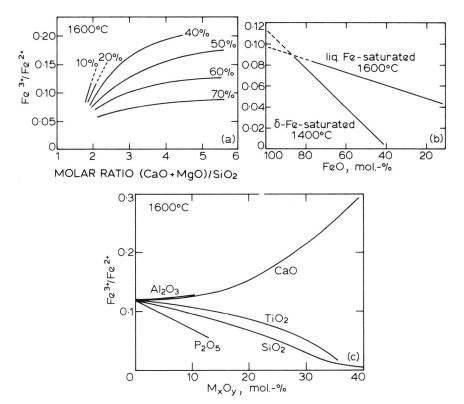

Fig. 5.29  Effect of slag composition on the ratio $Fe^{3+}/Fe^{2+}$ in melts saturated with metallic iron at 1600°C: (a) CaO–MgO–SiO$_2$–FeO melts at indicated molar concentrations of total iron oxide FeO; (b) MgO–SiO$_2$–FeO melts; (c) pseudobinary FeO–M$_x$O$_y$ melts. From Ref. 25.

Examples are given in Fig. 5.29 of variations of the ratio $Fe^{3+}/Fe^{2+}$ with slag composition of melts co-existing with liquid iron: (a) CaO–MgO–$SiO_2$–FeO, (b) MgO–$SiO_2$–FeO and (c) FeO–$M_xO_y$ melts, reproduced from a paper by Ban-ya and Shim.[25]

**5.7 SURFACE TENSION**

Surface tensions of liquid oxides and their mixtures are in the range 200 to 600 mN m$^{-1}$. In binary silicate melts, the surface tension decreases with an increasing silica content as shown in Fig. 5.30. Surface tensions measured by Kozakevitch[27] are given in Fig. 5.31 for binary melts with iron oxide, and in Fig. 5.32 for FeO–MnO–$SiO_2$ and FeO–CaO–$SiO_2$ melts at 1400°C. The surface tension of CaO–$Al_2O_3$–$SiO_2$ melts at 1600°C (with the mass ratio CaO/$Al_2O_3$ = 1) decreases from 670 to 530 mN m$^{-1}$ as the $SiO_2$ content increases from 0 to 40%.[28] The addition of CaS or $CaF_2$ to slags lowers the surface tension only slightly. For additional data on surface tensions of a wide variety of slags and fluxes, reference may be made to a review paper by Mills and Keene.[29]

5.7.1 SLAG-METAL INTERFACIAL TENSION

The slag–metal interfacial tensions have values between those for the gas-slag and gas-metal surface tensions. Consequently, the addition of

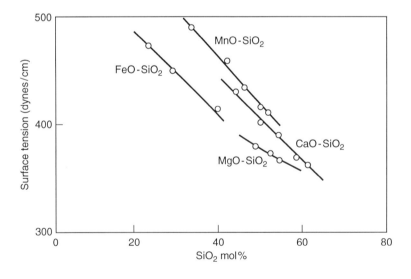

Fig. 5.30   Surface tensions of binary silicate melts: FeO–$SiO_2$ at 1420°C, other silicates at 1570°C. From Ref. 26.

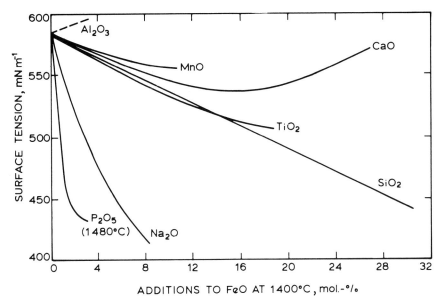

Fig. 5.31   Surface tensions of binary iron oxide melts at 1400°C. From Ref. 27.

surface active elements to liquid iron lowers the slag-metal interfacial tension.

Effects of sulphur and oxygen dissolved in iron on the interfacial tension between liquid iron and CaO–Al$_2$O$_3$–SiO$_2$ melts at 1600°C, determined by Gaye et al.[30], are shown in Figs. 5.33 and 5.34; compared to the surface tensions of Fe–S and Fe–O melts. The effect of oxygen on the interfacial tension is greater than sulphur, e.g. $\sigma_i$ = 600 mN m$^{-1}$ with $a_O$ = 0.05 ($\cong$ 0.05%) while at $a_S$ = 0.05 ($\cong$ 0.05%) $\sigma_i$ = 1000 mN m$^{-1}$.

As shown by Ogino *et al.*[31] (Fig. 5.35), a single curve describes adequately the effect of oxygen in iron on the interfacial tension between liquid iron and a wide variety of simple and complex slags, including those containing Na$_2$O and CaF$_2$. In the case of slags containing iron oxide, a decrease in interfacial tension with an increasing iron oxide content is due entirely to the corresponding increase in the oxygen content of the iron.

### 5.7.2 SOLUTE-INDUCED INTERFACIAL TURBULENCE

It has been known since the early studies of Thomson[32] and Marangoni[33] that the surface layer on an underlying liquid will move under the influence of a gradient of surface pressure, which is by definition the negative of the surface tension gradient. A temperature gradient or a solute concentration gradient in the plane of the surface (or interface) leading to unstable gradients of interfacial tension causes convection currents in the boundary layer, known as the **Marangoni effect**.

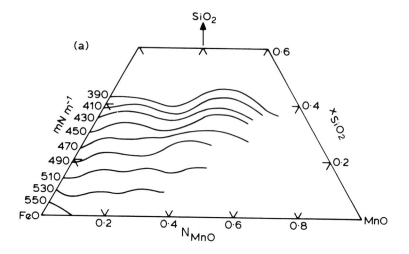

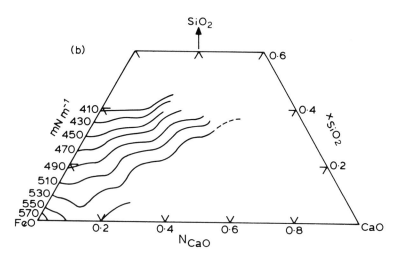

Fig. 5.32   Isosurface tension curves in FeO–MnO–SiO$_2$ and FeO–CaO–SiO$_2$ melts saturated with iron at 1400°C.

The phenomenon of interfacial turbulence accompanying mass transfer has been demonstrated in many investigations using either aqueous–organic solutions or gas–slag–metal systems. For a brief review of this subject, reference may be made to a previous publication.[10] However, a particular reference will be made here to the work of Riboud and Lucas[34] who measured changes in interfacial tension accompanying the transfer of alloying elements from liquid iron drops to aluminosilicate melts containing iron oxide. The interfacial tension was measured using the X-ray fluoroscopy technique. As is seen from an example of their experimental results in Fig. 5.36, the oxidation of aluminium in a liquid iron drop by the

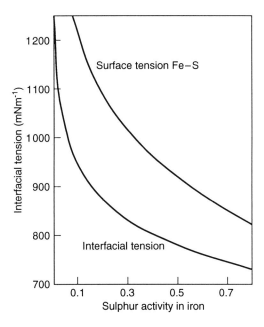

Fig. 5.33   Effect of sulphur in iron on the interfacial tension between liquid iron and CaO–Al$_2$O$_3$–SiO$_2$ melts, determined by Gaye *et al.*[30] is compared to the surface tension of Fe–S melts at 1600°C.

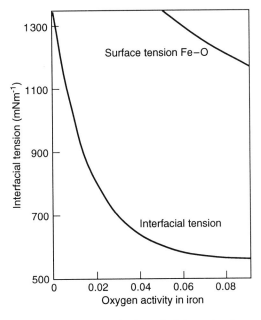

Fig. 5.34   Effect of oxygen in iron on the interfacial tension between liquid iron and CaO–Al$_2$O$_3$–SiO$_2$ melts, determined by Gaye *et al.*[30] is compared to the surface tension of Fe–O melt at 1600°C.

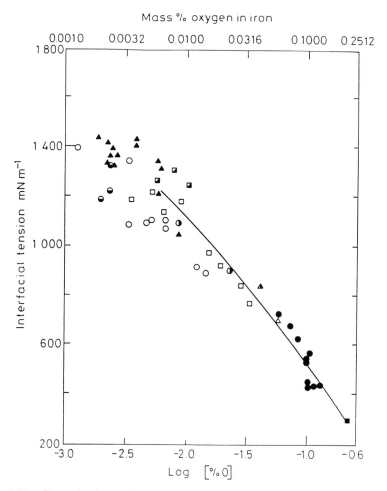

Fig. 5.35   General relation between the oxygen content of iron and the interfacial tension between the metal and various slag systems at 1580°C. From Ref. 31.

slag is accompanied by an appreciable decrease in the interfacial tension which reaches values close to zero. As the rate of transfer of aluminium from metal to slag decreases with the progress of the reaction, there is a restoring effect, i.e. the interfacial tension increases, ultimately reaching the equilibrium value of about 1200 mN m$^{-1}$ as $-d\%$ Al/d$t$ approaches zero.

## 5.8 DENSITY

Many repetitive measurements of slag densities have been made. Only selected references are given on the density data cited here for steelmaking type of slags. Since the density of silica (2.15 g cm$^{-3}$ at 1700°C) is much

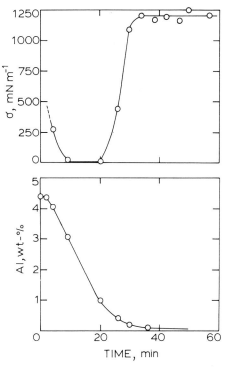

Fig. 5.36 Changes in metal–slag interfacial tension which occur during the transfer of aluminium from a liquid Fe–Al alloy drop to molten calcium silicate in an inert atmosphere. From Ref. 34.

lower than the densities of other metal oxide components of slags, densities of slags will decrease with an increasing silica content.

The density data for binary silicates are given in Fig. 5.37. The data in Fig. 5.38 are for CaO–MgO–$Al_2O_3$–$SiO_2$ melts, relevant to neutral slags for

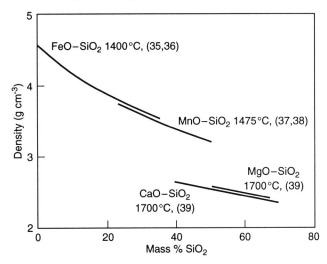

Fig. 5.37 Densities of binary silicate melt. From Refs. 35–39.

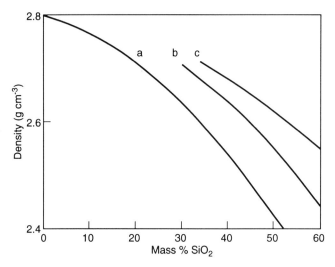

Fig. 5.38   Densities of CaO–MgO–Al$_2$O$_3$–SiO$_2$ melt at 1550°C, using data in Refs. 40–42. *a*: 0% MgO, CaO/Al$_2$O$_3$ = 1; *b*: 0% MgO, 5% Al$_2$O$_3$; *c*: 5% MgO, 5% Al$_2$O$_3$.

steel refining in the ladle. The density data in Fig. 5.39 compiled by Mills and Keene[29] are for simple and complex slags containing FeO, CaO, MgO, SiO$_2$ and P$_2$O$_5$. Since the densities of FeO–SiO$_2$ and MnO–SiO$_2$ are essentially the same, the average of the data in Fig. 5.39 is represented by the following equation in terms of (%FeO + %MnO).

$$\rho, \text{g cn}^{-3} = 2.46 + 0.018 \times (\%\text{FeO} + \%\text{MnO}) \tag{5.33}$$

### 5.9 VISCOSITY

The size of the silicate and aluminosilicate network in molten slags becomes larger with increasing SiO$_2$ and Al$_2$O$_3$ contents, hence their mobility decreases resulting in a higher viscosity. The addition of metal oxides or an increase in temperature leads to the breakdown of the Si(Al)O$_4$ network, resulting in lower melt viscosity.

Machin and Yee[43] made an extensive study of the viscosity of CaO–MgO–Al$_2$O$_3$–SiO$_2$ melts at temperatures of 1350 to 1500°C. The data in Fig. 5.40 is for the ternary system at 1500°C. The isokoms are approximately parallel to the binary side Al$_2$O$_3$–SiO$_2$, indicating that Al$_2$O$_3$ and SiO$_2$ are isomorphous in their effect on the slag viscosity. The isokoms in Fig. 5.41 are for the quaternary system with 35% and 50% SiO$_2$. In this case, the isokoms are approximately parallel to the binary side CaO–MgO, indicating that Ca$^{2+}$ and Mg$^{2+}$ cations have similar effects on the breakdown of the aluminosilicate network.

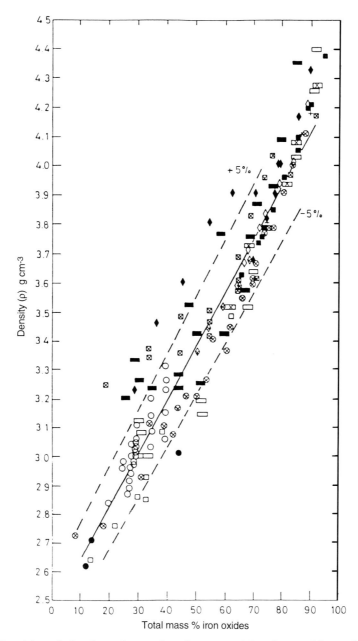

Fig. 5.39  Densities of simple and complex slags containing iron oxide at about 1400°C. From Ref. 29.

In an earlier study, Turkdogan and Bills[44] showed that for a given temperature the viscosity of CaO–MgO–Al$_2$O$_3$–SiO$_2$ melts is a single function of the mol fractions of SiO$_2$ and silica equivalence of Al$_2$O$_3$ as ($N_{SiO_2}$ + $N_a$). In slags containing less than 20% Al$_2$O$_3$, $N_a \approx N_{Al_2O_3}$. As is seen from

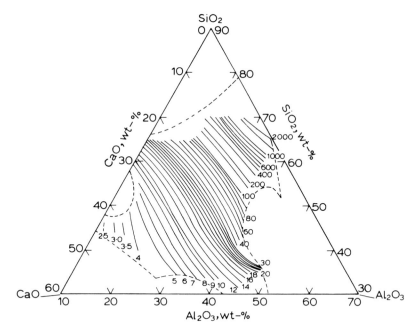

Fig. 5.40   Isokoms (0.1N s m$^{-2}$) for CaO–Al$_2$O$_3$–SiO$_2$ system at 1500°C. From Ref. 43.

the plot in Fig. 5.42, a single curve represents the viscosities of CaO–SiO$_2$[45], CaO–Al$_2$O$_3$–SiO$_2$[43,46] and CaO–MgO–Al$_2$O$_3$–SiO$_2$[43] melts. In a subsequent experimental work, Bills[47] found that this correlation also holds good for CaO–BaO–FeO–SiO$_2$ and CaO–MgO–FeO–Al$_2$O$_3$–SiO$_2$ melts.

For the temperature range 1400 to 1800°C, the temperature dependence of viscosity may be represented as follows for $(N_{SiO_2} + N_a) = 0.45$ and $0.65$

$$(N_{SiO_2} + N_a) = 0.45: \log \eta \ (0.1N \ s \ m^{-2}) = \frac{7477}{T} - 3.727 \qquad (5.34a)$$

$$(N_{SiO_2} + N_a) = 0.65: \log \eta \ (0.1N \ s \ m^{-2}) = \frac{10,425}{T} - 4.390 \qquad (5.34b)$$

As one would expect, the heat of activation for viscous flow increases with increasing contents of silica and alumina in the slag.

Viscosities of steelmaking slags are well represented by the experimental data of Kozakevitch[27] given in Fig. 5.43.

In the study of viscosities of mould fluxes for continuous casting, the experimental results have been represented as a function of temperature using the relation

$$\eta = AT \exp (B/T) \qquad (5.35)$$

where $A$ and $B$ are functions of slag composition. For the composition range (wt %) 33–56% SiO$_2$, 12–45% CaO, 0–11% Al$_2$O$_3$, O–20% Na$_2$O and

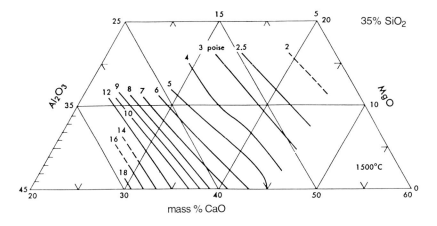

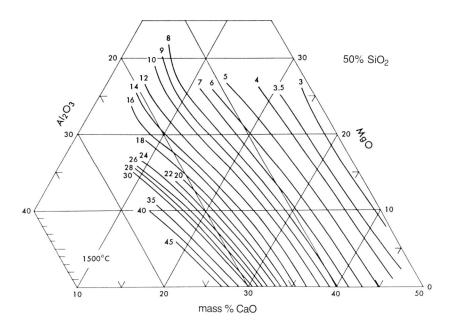

Fig. 5.41   Isokoms (0.1N s m⁻²) for CaO–MgO–Al₂O₃–SiO₂ system at 1500°C for melts containing 35% and 50% SiO₂. From Ref. 43.

–20% CaF₂, an interpolation formula has been derived for the parameters $A$ and $B$ as a function of the mole fractions of the constituents as given below.[48]

$$\ln A = -17.51 - 35.76 \times (Al_2O_3) + 1.73 \times (CaO) \qquad (5.36)$$
$$+ 5.82 \times (CaF_2) + 7.02 \times (Na_2O)$$

$$B = 31140 + 68833 \times (Al_2O_3) - 23896 \times (CaO) \qquad (5.37)$$
$$- 46351 \times (CaF_2) - 39519 \times (Na_2O)$$

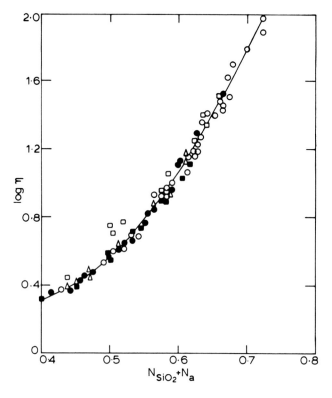

Fig. 5.42   Viscosities of CaO–MgO–Al$_2$O$_3$–SiO$_2$ melts at 1500°C as a function of SiO$_2$ and Al$_2$O$_3$ contents. Points ○ ● ■ from Ref. 43; Points △ from Ref. 45; points □ from Refs. 43 and 46.

where $A$ is in units of 0.1 N s m$^{-2}$K$^{-1}$ ($\equiv$poise/deg.) and $B$ is in degrees Kelvin.

### 5.10 MASS DIFFUSIVITY

Because of the ionic nature of molten slags, the diffusive mass transfer is by ions. The ionic diffusivities are measured using the radioactive tracer elements dissolved in an oxidised form in the melt. Typical examples of ionic diffusivities in slags at 1600°C are given below.

| Ion | $D_i^*$, cm$^2$/s |
|---|---|
| Si$^{4+}$, O$^{2-}$ | $4 \times 10^{-7} - 1 \times 10^{-6}$ |
| Al$^{3+}$ | $\approx 1 \times 10^{-7}$ |
| Ca$^{2+}$, Mg$^{2+}$, Fe$^{2+}$ | $6 \times 10^{-6} - 1 \times 10^{-5}$ |
| S$^{-2}$ | $\approx 4 \times 10^{-6}$ |

Since the electro neutrality has to be maintained, diffusion of a cation is

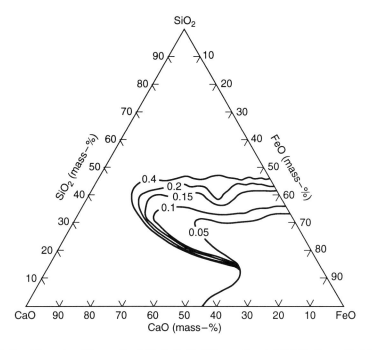

Fig. 5.43   Viscosity (N s m⁻²) of CaO–FeO–SiO₂ melts at 1400°C. From Ref. 27.

accompanied by diffusion of the oxygen ion. The diffusion that occurs in the dissolution of a solid oxide in the slag is controlled by the mobility of the $O^{2-}$ ion which is smaller than the divalent cations.

## 5.11 ELECTRICAL CONDUCTIVITY

The electrical current in molten slags is carried by the cations. However, in slags containing high concentrations of FeO or MnO (>70%) the electronic conduction becomes the dominant mechanism.

The ionic conductivity $\lambda_i$ is theoretically related to the self diffusivity of the ionic species i by the Nernst–Einstein equation

$$D^*_i = \frac{RT}{F^2 Z_i^2 C_i} \lambda_i \qquad (5.38)$$

where  $F$  = Faraday constant, 96,500 C mol⁻¹
$Z_i$ = valency of ion $i$
$C_i$ = concentration of ion $i$, mol cm⁻³
$\lambda_i$ = specific conductivity, $\Omega^{-1}$cm⁻¹
For steelmaking slags:   $\lambda = 0.5 - 1.5\ \Omega^{-1}$cm⁻¹
For ladle slags:          $\lambda = 0.4 - 0.7\ \Omega^{-1}$cm⁻¹

The electrical conductivity increases with an increasing slag basicity and increasing temperature.

## 5.12 THERMAL CONDUCTIVITY

Because of the presence of iron oxide, the metallurgical slags are opaque to infrared radiation, therefore the heat conduction is primarily thermal.

Thermal conductivity of slags and mould fluxes are in the range 0.5 to 1.2 $Wm^{-1}K^{-1}$. From experimental data the following approximate empirical relation has been found

$$\kappa(Wm^{-1}K^{-1}) = 1.8 \times 10^{-5}V^{-1} \qquad (5.39)$$

where V is the molar volume $= \dfrac{M}{\rho}$, $m^3mol^{-1}$.

## REFERENCES

1. E.M. LEVIN, C.R. ROBBINS and H.F. MCMURDIE, *Phase diagrams for ceramists*, 1974 3rd edn. Am. Ceram. Soc.
2. M.L. KEITH and O.F. TUTTLE, *Am. J. Sci*, 1952, 'Bowen Volume', Part I, 203.
3. O.W. FLORKE, *Naturwissen*, 1956, **43**, 419.
4. S.B. HOLMQUIST, *J. Am. Ceram. Soc.*, 1961, **44**, 82.
5. G. TROMEL, K. KOCH, W. FIX and N. GROBKURTH, *Arch. Eisenhüttenwes.*, 1969, **40**, 969.
6. K.L. FETTERS and J. CHIPMAN, *Trans. AIME*, 1941, **145**, 95.
7. R.J. LEONARD and R.H. HERRON, *NOH–BOS Conference Proc.*, 1977, **60**, 127.
8. R. SELIN, *Scan J. Metallurgy*, 1991, **20**, 279.
9. E. GORL, R. KLAGES, R. SCHEEL and G. TROMEL, *Arch Eisenhüttenwes.*, 1969, **40**, 959.
10. E.T. TURKDOGAN, *Physicochemical properties of molten slags and glasses*, The Metals Society (now The Institute of Materials), London, 1983.
11. C.R. TAYLOR and J. CHIPMAN, *Trans. AIME*, 1943, **159**, 228.
12. M. TIMUCIN and A.E. MORRIS, *Metall. Trans*, 1970, **1**, 3193.
13. R.H. REIN and J. CHIPMAN, *Trans. Met. Soc. AIME*, 1965, **233**, 415.
14. T. FUJISAWA and H. SAKAO, *Tetsu-to-Hagané*, 1977, **63**, 1494, 1504.
15. T.B. WINKLER and J. CHIPMAN, *Trans. AIME*, 1946, **167**, 111.
16. H. KNÜPPEL and F. OETERS, *Stahl u. Eisen*, 1961, **81**, 1437.
17. G. TROMEL and W. FIX, *Arch. Eisenhüttenwes*, 1962, **33**, 745.
18. H.B. BELL, *J. Iron and Steel Inst.*, 1963, **201**, 116.
19. H. SUITO and R. INOUE, *Trans. Iron and Steel Inst. Japan*, 1984, **24**, 257.
20. T. USUI, K. YAMADA, Y. KAWAI, S. INOUE, H. ISHIKAWA and Y. NIMURA, *Tetsu-to-Hagané*, 1991, **77**, 1641.
21. E.T. TURKDOGAN, *Ironmaking and Steelmaking*, 1993, **20**, 469.
22. F.D. RICHARDSON and J.H.E. JEFFES, *J. Iron and Steel Inst.*, 1948, **160**, 261.

23. Y. IGUCHI, S. BAN-YA and T. FUWA, *Trans. Iron and Steel Inst, Japan*, 1969, **9**, 189.
24. K. ITO and R.J. FRUEHAN, *Metall. Trans. B*, 1988, , **19B**, 419.
25. S. BAN-YA and J.–D. SHIM, *Can. Metall. Q.*, 1982, **21**, 319.
26. T.B. KING, *J. Soc. Glass Technol.*, 1951, **35**, 241.
27. P. KOZAKEVITCH, *Rev. Métall.*, 1949, **46**, 505, 572.
28. A.S. KIM, A.A. AKBERDIN and I.S. KULIKOV, VINITI (USSR) Rept. No. 5508–81, 1981.
29. K.C. MILLS and B.J. KEENE, *Inter. Met. Rev.*, 1981, **26**, 21.
30. H. GAYE, L.D. LUCAS, M. OLETTE and P.V. RIBOUD, *Can. Metall. Q.*, 1984, **23**, 179.
31. K. OGINO, S. HARA, T. MIWA and S. KIMOTO, *Trans. Iron and Steel Inst. Japan.*, 1984, **24**, 522.
32. J.J. THOMSON, *Philos. Mag.*, 1855, **10**(5), 330.
33. C. MARANGONI, *Ann. Phys. Chem.*, 1871, **143**, 337.
34. P.V. RIBOUD and L.D. LUCAS, *Can. Metall. Q.*, 1981, **20**, 199.
35. D.R. GASKELL, A. MCLEAN and R.G. WARD, *Trans. Farady Soc.*, 1969, **65**, 1498.
36. Y. SKIRAISHI, K. IKEDA, A. TAMURA and T. SAITO, *Trans. Japan Inst. Met.*, 1978, **19**, 264.
37. YU M. GOGIBERIDZE, M.A. KEKELIDZE and SH. M. MIKIASHVILI, *Soob A Gruz. SSR*, 1963, **32**(1), 117.
38. L. SEGERS, A. FONTANA and R. WINAND, *Electrochem. Acta.*, 1978, **23**, 1275.
39. J.W. TOMLINSON, M.S.R. HEYNES and J.O'M. BOCKRIS, *Trans. Faraday Soc.*, 1958, **54**, 1822.
40. L.R. BARRETT and A.G. THOMAS, *J. Glas Technol.*, 1959, **43**, 179s.
41. H. WINTERHAGER, L. GREINER and R. KAMMEL, *Forschungsberichte des Landes, Nordrhein-Westfalen*, Nr. 1630. Westdeutscher Verlag, 1966.
42. E.V. KRINOCHKIN, K.T. KUROCHIN and P.V. UMRIKHIN, *Fiz. Khim. Poverkh. Yavlenii Rasp.*, 1971, pp. 179–83. Naukova Dumka, Kiev.
43. J.S. MACHIN and T.B. YEE, *J. Am. Cerm. Soc.*, 1948, **31**, 200; 1954, **37**, 177.
44. E.T. TURKDOGAN and P.M. BILLS, *Am. Ceram. Soc.*, 1960, **39**, 682.
45. J. O'M. BOKRIS and D.C. LOWE, *Proc. R. Soc.*, 1954, **226A**, 423.
46. P. KOZAKEVITCH, *Rev. Métall.*, 1960, **57**, 149.
47. P.M. BILLS, *J. Iron and Steel Inst.*, 1963, **201**, 133.
48. P.V. RIBOUD, M. OLETTE, J. LECREC and W. POLLOK, *NOH–BOS Conf. Proc.*, 1978, **61**, 411.

# CHAPTER 6

# Equilibrium Data on Liquid Steel–Slag Reactions

From the physicochemical properties of liquid iron alloys and slags documented in Chapters 4 and 5 and the available experimental data on slag–metal reactions, an assessment will be made of the equilibrium states of slag–metal reactions and deoxidation reactions pertinent to steelmaking and steel refining in the ladle.

Understanding the physics and chemistry of ironmaking and steelmaking reactions has been the subject of many repetitive studies since the early 1930s, both in academia and in government and industry sponsored research laboratories. The manner of applying the principles of thermodynamics and physical chemistry has gone through many different phases of development, searching for a better understanding of the operation and control of the pyrometallurgical processes. Reflecting upon the many facets of these previous endeavours, the author has now reached the conclusion that the equilibrium states of slag–metal reactions, pertinent to the steelmaking conditions, can be quantified in simpler forms. It is the author's considered opinion that the equilibrium constants of slag–metal reactions vary with the slag composition in different ways, depending on the type of reaction. For some reactions the slag basicity is the key parameter to be considered; for another reaction the key parameter could be the mass concentration of either the acidic or basic components of the slag. The equilibrium relations given are for low alloy steels in which the activities of dissolved elements are essentially equivalent to their mass concentrations.

## 6.1 OXIDATION OF IRON

In steelmaking slags, the total number of g-mols of oxides per 100 g of slag is within the range $1.65 \pm 0.05$. Therefore, the analysis of the slag–metal equilibrium data, in terms of the activity and mol fraction of iron oxide given in section 5.4.4, can be transposed to a simple relation between the mass ratio [ppm O]/(%FeO) and the sum of the acidic oxides $\%SiO_2 + 0.84 \times \%P_2O_5$ as depicted in Fig. 6.1a for slag basicities of $B>2$. The experimental data used in the diagram are those cited in Fig. 5.22. There is of course a corollary relation between the ratio [ppm O]/(%FeO) and the slag basicity as shown in Fig. 6.1b.

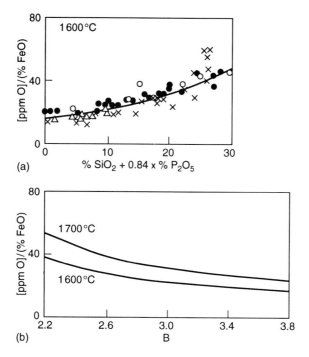

(a)

(b)

Fig. 6.1   Equilibrium ratio [ppm O]/(%FeO) related to SiO₂ and P₂O₅ contents and slag basicity; experimental data are those cited in Fig. 5.22.

## 6.2 OXIDATION OF MANGANESE

For the FeO and MnO exchange reaction involving the oxidation of manganese in steel formulated below

$$(FeO) + [Mn] = (MnO) + [Fe] \qquad (6.1)$$

the equilibrium relation may be described in terms of the mass concentrations of oxides

$$K'_{FeMn} = \frac{(\%MnO)}{(\%FeO)\,[\%Mn]} \qquad (6.2)$$

where the equilibrium relation $K'_{FeMn}$ depends on temperature and slag composition. The values of $K'_{FeMn}$, derived from the equilibrium constant $K_{FeMn}$ in equation (5.13b) and the activity coefficient ratios $\gamma_{FeO}/\gamma_{MnO}$ in Fig. 5.23, are plotted in Fig. 6.2 against the slag basicity. In BOF, Q–BOP and EAF steelmaking, the slag basicities are in the range 2.5 to 4.0 and the melt temperature in the vessel at the time of furnace tapping in most practices is between 1600–1650°C, for which the equilibrium $K'_{FeMn}$ is about 1.8 ± 0.2.

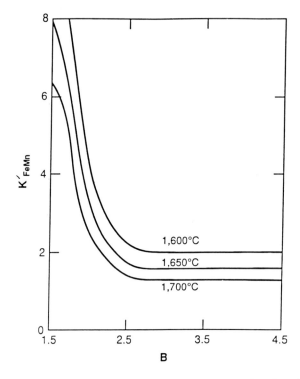

Fig. 6.2   Equilibrium relation $K'_{\text{FeMn}}$ in equation (6.2) related to slag basicity.

### 6.3 OXIDATION OF CARBON

As noted in section 4.2.4, for the reaction

$$CO(g) = [C] + [O] \tag{6.3}$$

the equilibrium constant for low alloy steels containing less than 1% C for steelmaking temperatures, is

$$K = \frac{[\%C][\text{ppm O}]}{p_{CO} \text{ (atm)}} = 20 \tag{6.4}$$

With respect to the slag-metal reaction, the equilibrium relation for carbon oxidation would be

$$(FeO) + [C] = CO + [Fe] \tag{6.5}$$

$$K_{FC} = \frac{p_{CO} \text{ (atm)}}{[\%C]a_{FeO}} \tag{6.5a}$$

$$\log K_{FC} = -\frac{5730}{T} + 5.096 \tag{6.5b}$$

For 1600°C, $\gamma_{FeO}$ = 1.3 at slag basicity of $B$ = 3.2 and $p_{CO}$ = 1.5 atm (average CO pressure in the vessel), we obtain the following equilibrium relation between the carbon content of steel and the iron oxide content of slag.

$$K_{FC} = 108.8$$

$$a_{FeO} = 1.3N_{FeO} \approx \frac{1.3}{72 \times 1.65} \; (\%FeO) = 0.11 \times (\%FeO)$$

$$(\%FeO)\,[\%C] = 1.25 \tag{6.6}$$

For the reaction

$$(MnO) + [C] = CO + [Mn] \tag{6.7}$$

the following equilibrium relations apply

$$K_{MC} = \frac{p_{CO}\,(atm)\,[\%Mn]}{[\%C]a_{MnO}} \tag{6.7a}$$

$$\log K_{MC} = -\frac{13{,}182}{T} + 8.574 \tag{6.7b}$$

$K_{MC}$ = 34.4 at 1600°C, $\gamma_{MnO}$ = 2.05 at $B$ = 3.2 and $a_{MnO} = \frac{2.05}{71 \times 1.65}\,(\%MnO)$ = 0.017 $(\%MnO)$. For these values and 1.5 atm CO we obtain the following equilibrium relation.

$$\frac{[\%Mn]}{(\%MnO)} = 0.4\,[\%C] \tag{6.8}$$

## 6.4 OXIDATION OF SILICON

For low silicon contents we may use mass concentrations of Si and O in the equilibrium constant for the reaction

$$[Si] + 2[O] = (SiO_2) \tag{6.9}$$

$$K_{Si} = \frac{a_{SiO_2}}{[\%Si][\%O]^2} \tag{6.9a}$$

where the silica activity is with respect to solid $SiO_2$ as the standard state. The temperature dependence of $K_{Si}$ is given by

$$\log K_{Si} = \frac{30{,}410}{T} - 11.59 \tag{6.9b}$$

For the BOF tap temperatures $K \approx 3 \times 10^4$ and for lime-saturated slags $a_{SiO_2}$ < 0.01; therefore for 800 ppm O in steel at turndown the equilibrium content of silicon in the steel would be less than 1 ppm Si. In practice the steel contains 0.003 to 0.005% Si at turndown. Evidently, at these low concentrations of silicon the reaction kinetics are no longer favourable to reach slag–metal equilibrium with respect to silicon oxidation.

## 6.5 OXIDATION OF CHROMIUM

There are two valencies of chromium ($Cr^{2+}$ and $Cr^{3+}$) dissolved in the slag. The ratio $Cr^{2+}/Cr^{3+}$ increases with an increasing temperature, decreasing oxygen potential and decreasing slag basicity. Under steelmaking conditions, i.e. in the basic slags and at high oxygen potentials, the trivalent chromium predominates in the slag. The equilibrium distribution of chromium between slag and metal for basic steelmaking slags, determined by various investigators,[1–3] is shown in Fig. 6.3; slope of the line represents an average of these data.

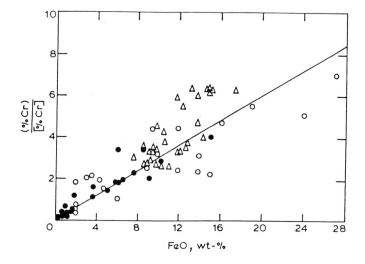

Fig. 6.3   Variation of chromium distribution ratio with the iron oxide content of slag, in the ($\Delta$) open hearth[1] and ($\circ$) electric arc furnace[3] at tap is compared with the results of laboratory experiments[2] ($\bullet$).

$$\frac{(\%Cr)}{[\%Cr]} = (0.3 \pm 0.1) \times (\%FeO) \qquad (6.10)$$

In the AOD stainless steelmaking with bottom blowing argon–oxygen mixtures, the chromium and carbon are oxidised independently of each other. As the chromic oxide particles float out of the melt together with gas bubbles they react with carbon, resulting in some chromium recovery back into the steel. The temperature and gas composition have a significant effect on the direction of the reaction.

$$Cr_2O_3(s) + 3[C] \rightleftarrows 2[Cr] + 3CO \ (g) \tag{6.11}$$

The equilibrium relations for the above reaction, determined experimentally by Richardson and Dennis,[4] are plotted in Fig. 6.4 for Fe–Cr–C melts saturated with $Cr_2O_3$.

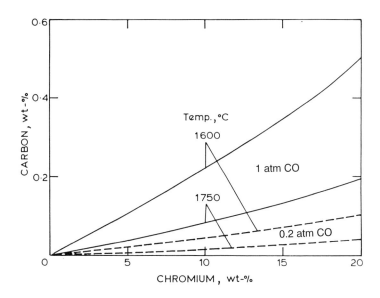

Fig. 6.4  Carbon–chromium relation in liquid steel in equilibrium with solid $Cr_2O_3$ at indicated temperatures and CO pressures. From Ref. 4.

By lowering the partial pressure of CO in gas bubbles with an argon dilution of the oxygen blow, and operating the furnace at high temperatures, the steel can be decarburised to low levels with a minimum oxidation of chromium. As the concentration of carbon decreases, the $O_2/Ar$ ratio in the blow is decreased to suppress an excessive oxidation of chromium.

When steel is decarburised to the desired level, usually < 0.05% C, about 3% Cr in the Steel is oxidised despite close control of the $O_2/Ar$ ratio in the blow. Chromium is recovered from the slag by reducing its oxide with silicon. Under this reducing condition the reaction to be considered is as formulated below.

$$2(CrO) + [Si] = 2[Cr] + (SiO_2) \tag{6.12}$$

The typical compositions of AOD slags after decarburisation and after silicon reduction are given in Table 6.1. As is seen from the experimental data in Fig. 6.5, the higher the slag basicity and higher the silicon content of steel, the lower is the equilibrium slag/metal distribution of chromium, i.e. the greater the chromium recovery from the slag.

**Table 6.1**   Ranges of AOD slag composition after decarburisation and after silicon reduction

|  | Composition, wt.% | |
|---|---|---|
|  | after decarburisation | after silicon reduction |
| FeO | 4–6 | 1–2 |
| MnO | 4–8 | 1–3 |
| $SiO_2$ | 12–18 | 30–40 |
| $Al_2O_3$ | 18–22 | 3–8 |
| CaO | 8–15 | 33–43 |
| MgO | 7–15 | 10–20 |
| $Cr_3O_4$* | 20–30 | 1–3 |

* Incorrect formulation often reported in plant data.

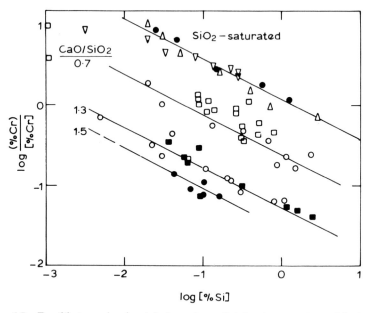

Fig. 6.5   Equilibrium slag/metal chromium distribution varying with the concentration of silicon in iron coexisting with chromium oxide containing $CaO–Al_2O_3–SiO_2$ slags at temperatures of 1600° to 1690°C. See Ref. 5 for references to experimental data.

## 6.6 OXIDATION OF PHOSPHORUS

The state of equilibrium of the phosphorus reaction between slag and low alloy steel has been described and formulated in a variety of ways during the past six decades. The incorrect earlier formulations are shown below.

$$3(CaO) + 5(FeO) + 2[P] = (Ca_3P_2O_8) \tag{6.13}$$

$$2[P] + 5[O] = (P_2O_5) \tag{6.14}$$

It was in the late 1960s that the correct formulation of the phosphorus reaction was at last realised, thus

$$[P] + \tfrac{5}{2}[O] + \tfrac{3}{2}(O^{2-}) = (PO_4^{3-}) \qquad (6.15)$$

At low concentrations of [P] and [O], as in most of the experimental melts, their activity coefficients are close to unity, therefore mass concentrations can be used in formulating the equilibrium relation $k_{PO}$ for the above reaction.

$$k_{PO} = \frac{(\%P)}{[\%P]} \; [\%O]^{-2.5} \qquad (6.16)$$

The equilibrium relation $k_{PO}$, known as the phosphate capacity of the slag, depends on temperature and slag composition. The values of $k_{PO}$ are derived from the experimental data of several independent studies,[6–10] as discussed below.

In the experiments of Nagabayashi *et al.*[8] the MgO-saturated FeO–MgO–$P_2O_5$ slags were equilibrated with liquid iron containing 1.5 to 3.0% P and 0.1 to 0.2% O. For these high solute concentrations, the phosphorus and oxygen interaction coefficients have to be incorporated in the calculation of $k_{PO}$ using the equations below.

$$\log f_P = 0.062[\%P] \qquad (6.17a)$$

$$\log f_O = 0.07[\%P] - 0.1[\%O] \qquad (6.17b)$$

As is seen from the experimental data plotted in Fig. 6.6, the slag basicity $B$ is not a suitable parameter to describe the composition

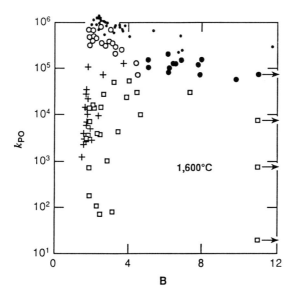

Fig. 6.6   Phosphate capacities of slags do not relate to basicity.

dependence of the phosophate capacity of slags. Of the basic oxides present in the steelmaking type of slags, it is primarily %CaO (+ $CaF_2$) and secondarily %MgO in the slag which strongly influence the phosphate capacity of the slag.

By comparing the $k_{PO}$ values for the system $FeO–MgO–P_2O_5$ with those for the lime containing slags, %CaO equivalence of MgO is estimated to be $0.3 \times$ %MgO. The %$CaF_2$ is taken to be equivalent to %CaO. On this premise, the sum of these basic oxides as

$$BO = \%CaO + \%CaF_2 + 0.3 \times \%MgO \qquad (6.18)$$

is considered to be the key parameter in describing the composition dependence of the phosphate capacity of the slag. The $k_{PO}$ values of widely different slag compositions are plotted in Fig. 6.7.

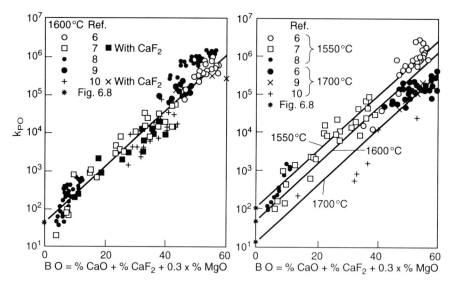

Fig. 6.7   Equilibrium data showing the decisive effects of CaO, $CaF_2$ and MgO on phosphate capacities of simple and complex slags.

The observed scatter in the $k_{PO}$ values is random and not related in any systematic manner to variations in the concentrations of other oxide components of the slag. The average $k_{PO}$ values are represented by the parallel lines drawn to intersect the ordinate at points * for pure iron phosphate melts.

Several studies were made in the 1930s[11–14] and subsequently by Tromel and Schwerdtfeger[15] of the phosphorus reaction between iron phosphate slags and liquid iron containing up to about 10% P. About ten years ago, Ban-ya *et al.*[16] measured the activity of iron oxide in iron phosphate melts at temperatures of 1200 to 1450°C. In deriving the values of $k_{PO}$ from these experimental data, due account was taken of the effect of phosphorus in iron on the activity coefficients $f_P$ and $f_O$. Also in the phosphate melts containing MgO, the $k_{PO}$ values were extrapolated to zero MgO. The results are given in Fig. 6.8.

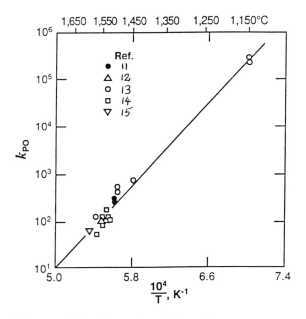

Fig. 6.8    Phosphate capacities of iron phosphate melts.

From the foregoing critical reassessment of the slag–metal equilibrium data of several independent studies, the following equation is derived to describe the effects of temperature and slag composition on the phosphate capacities of steelmaking type of slags.

$$\log k_{PO} = \frac{21740}{T} - 9.87 + 0.071 \times BO \qquad (6.19)$$

## 6.7 REDUCTION OF SULPHUR

The sulphur transfer from metal to slag is a reduction process as represented by this equation

$$[S] + (O^{2-}) = (S^{2-}) + [O] \qquad (6.20)$$

for which the state of slag-metal equilibrium is represented by

$$k_{SO} = \frac{(\%S)}{[\%S]} [\%O] \qquad (6.20a)$$

As is seen from the plots in Fig. 6.9, the sulphide capacities of slags, $k_{SO}$, measured in three independent studies are in general accord. The effect of temperature on $k_{SO}$ is masked by the scatter in the data. The concentration of acidic oxides, e.g. $\%SiO_2 + 0.84 \times \%P_2O_5$, rather than the slag basicity seems to be a better representation of the dependence of $k_{SO}$ on the slag composition.

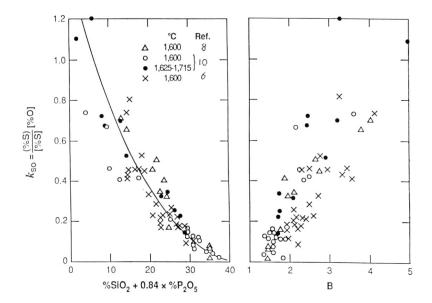

Fig. 6.9   Sulphide capacities of slags.

In view of the relationship between the ratio $[\text{ppm O}]/(\%FeO)$ and the sum of the acidic oxides in Fig. 6.1, the sulphide capacity of the slag may be represented also by the following expression.

$$k_S = \frac{(\%S)}{[\%S]}\,(\%FeO] \qquad (6.21)$$

As the concentrations of $SiO_2$ and $P_2O_5$ increase, the value of $k_S$ decreases, as shown in Fig. 6.10 which is reproduced from a previous publication.[5]

For steel desulphurisation in the ladle with the calcium aluminate-based slag and aluminium addition to steel, the slag–metal equilibrium data are for the reaction given below

$$\tfrac{2}{3}[Al] + [S] + (O^{2-}) = (S^{2-}) + \tfrac{1}{3}(Al_2O_3) \qquad (6.22)$$

for which the equilibrium relation is

$$K_{SA} = \frac{(\%S)}{[\%S]}\,[\%Al]^{-2/3} \qquad (6.22a)$$

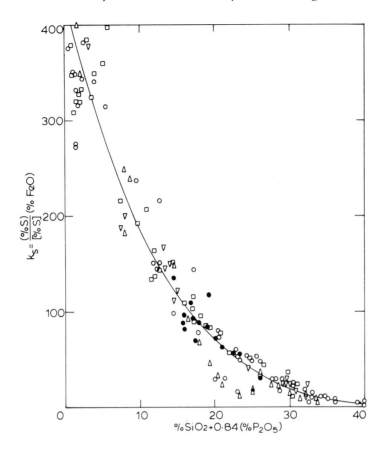

Fig. 6.10   Variation of the equlibrium constants $K_S$ with $SiO_2$ and $P_2O_5$ contents of simple and complex slags at temperatures of 1535° to 1700°C. From Ref. 5.

Numerous experimental studies have been made of the sulphide capacities of simple and complex slags by many investigators. The available gas–slag–metal equilibrium data, compiled by the author[17] in a critical review of the subject, are used to derive the equilibrium relations pertinent to steel desulphurisation in the ladle.

The effects of temperature and composition of calcium aluminate melts on the equilibrium constant $K_{SA}$ are shown in Fig. 6.11. As is seen from the reassessed equilibrium data in Fig. 6.12 for 1600°C, $K_{SA}$ decreases with an increasing $SiO_2$ content but increases with an increasing MgO content of the slag.

The solubilities of CaS in $CaO–Al_2O_3$ melts have been measured (Fig. 6.13). At 1600°C and lime saturation the solubility is 4.77% CaS ≡ 2.12% S. If the sulphur content of the steel is high and the slag volume is low, the ladle slag will become saturated with CaS during desulphurisation. As the desulphurisation continues the excess CaS formed becomes dispersed in the slag, resulting in a decrease in the dissolved CaO content of the slag, hence decreasing the value of $K_{SA}$.

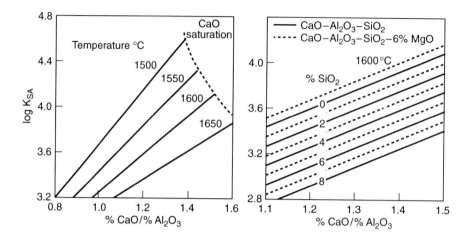

Fig. 6.11   Effects of temperature and composition of calcium aluminate melts on the quilibrium constant $K_{SA} = \dfrac{(\%S)}{[\%S]}$ $[\%Al]^{-2/3}$.

Fig. 6.12   Effects of $SiO_2$ and MgO on the equilibrium constant $K_{SA}$ at 1600°C.

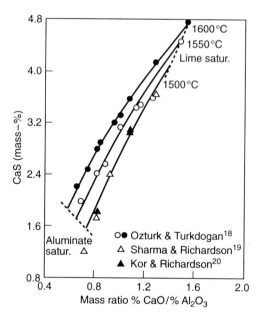

Fig. 6.13   Solubility of CaS in calcium aluminate melts at indicated temperatures as a function of mass ratio $\%CaO/\%Al_2O_3$.

## 6.8 Deoxidation Reaction Equilibrium

There are primarily three elements used in steel deoxidation:

Mn  $\Big\{$  as low and high C ferro alloy
Si    or as silicomanganese alloy
Al     about 98% purity

### 6.8.1 Deoxidation with Fe/Mn

When the steel is partially deoxidised with Mn, the iron also participates in the reaction, forming liquid or solid Mn(Fe)O as the deoxidation product.

$$\left. \begin{array}{l} [Mn] + [O] \rightarrow MnO \\ [Fe] + [O] \rightarrow FeO \end{array} \right\} \text{ liquid or solid Mn(Fe)O} \qquad (6.23)$$

The state of equilibrium of steel with the deoxidation product Mn(Fe)O is shown in Fig. 6.14.

### 6.8.2 Deoxidation with Si/Mn

Depending on the concentrations of Si and Mn added to steel in the tap

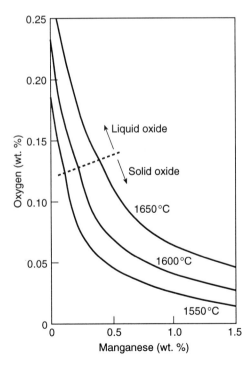

Fig. 6.14   Manganese and oxygen contents of iron in equilibrium with FeO–MnO liquid or solid solution.

ladle, the deoxidation product will be either molten manganese silicate or
solid silica.

$$[Si] + 2[O] \rightarrow SiO_2 \quad \left.\right\} \quad \text{molten } xMnO \cdot SiO_2$$
$$[Mn] + [O] \rightarrow MnO \quad \text{or solid } SiO_2 \qquad (6.24)$$

One of the early pioneering studies of slag–metal reaction equilibria is
that attributed to Korber and Oelsen[21] for their measurement of the
equilibrium distribution of manganese and silicon between liquid iron and
MnO–FeO–SiO$_2$ slags saturated with silica. The results of their experiments
at $1600 \pm 10°C$ are shown in Fig. 6.15 by way of example; these equilibrium
data have been substantiated in numerous subsequent studies.

Value of the equilibrium constant for Si deoxidation is already given in
equation (6.9). The following equilibrium relation is obtained for the Si/
Mn deoxidation reaction

$$[Si] + 2(MnO) = 2[Mn] + (SiO_2) \qquad (6.25)$$

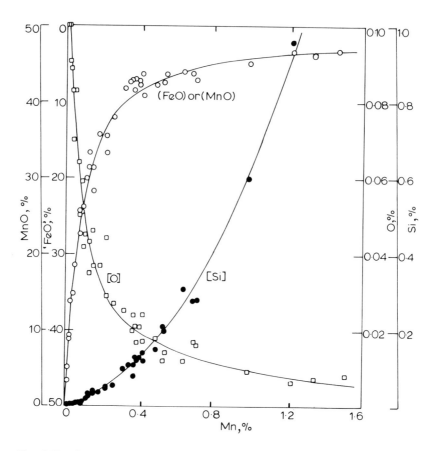

Fig. 6.15   Concentrations of Mn, Si and O in liquid iron equilibrated with SiO$_2$-
saturated iron-manganese silicate melts at $1600° \pm 10°C$. From Ref. 21.

$$K_{MnSi} = \left\{ \frac{[\%Mn]}{a_{MnO}} \right\}^2 \frac{a_{SiO_2}}{[\%Si]} \qquad (6.25a)$$

$$\log K = \frac{1510}{T} + 1.27 \qquad (6.25b)$$

where the oxide activities are relative to pure solid oxides. For high concentrations of silicon (>0.4%) the activity coefficient $f_{Si}$ should be used in the above equation, thus $\log f_{Si} = 0.11 \times [\%Si]$.

The activities of MnO in manganese silicate melts have been measured

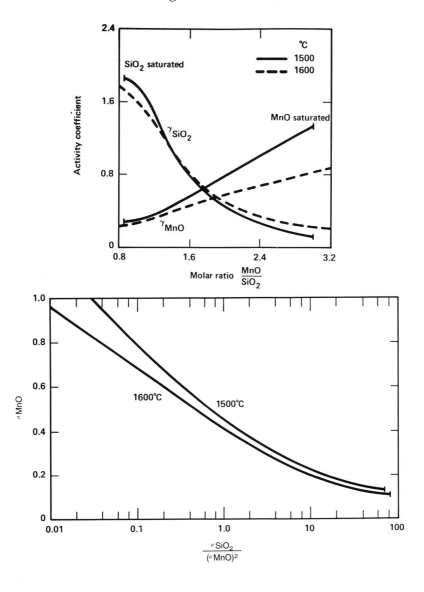

Fig. 6.16  Activities in MnO–SiO$_2$ melts with respect to solid oxides, derived from the experimental data in Refs. 22 and 23.

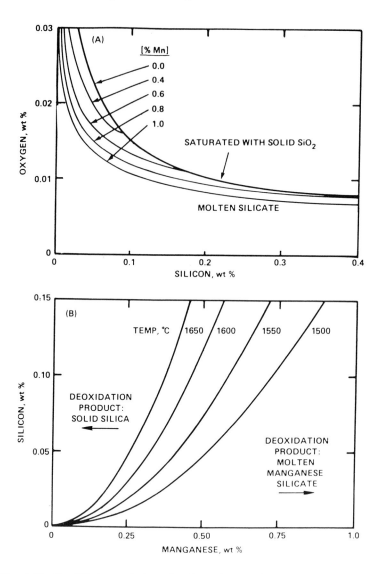

Fig. 6.17 Equilibrium relations for deoxidation of steel with silicon and manganese at 1600°C.

by Rao and Gaskell.[22] Their results are in substantial agreement with the results of the earlier work by Abraham *et al.*[23] The activity coefficients of the oxides (relative to solid oxides) are plotted in Fig. 6.16. For liquid steel containing Mn>0.4% the deoxidation product is a MnO-rich silicate with FeO<8%; therefore the activity data in Fig. 6.16 can be used together with equations (6.9) and (6.25) in computing the equilibrium state of the Si/Mn deoxidation as given in Fig. 6.17a. The deoxidation product being either solid silica or molten manganese silicate depends on temperature, Si and Mn contents, as shown in Fig. 6.17b.

### 6.8.3 DEOXIDATION WITH Si/Mn/Al

Semi-killed steels with residual dissolved oxygen in the range 40 to 25 ppm are made by deoxidising steel in the tap ladle with the addition of a small amount of aluminium together with silicomanganese, or a combination of ferrosilicon and ferromanganese. In this case, the deoxidation product is molten manganese aluminosilicate having a composition similar to $3MnO \cdot Al_2O_3 \cdot 3SiO_2$. With a small addition of aluminium, e.g. about 35 kg for a 220 to 240 t heat together with Si/Mn, almost all the aluminium is consumed in this combined deoxidation with Si and Mn. The residual dissolved aluminium in the steel will be less than 10 ppm. As is seen from Fig. 5.21, for the deoxidation product $MnO–Al_2O_3–SiO_2$ saturated with $Al_2O_3$, the silica activities are 0.27 at 1650°C, 0.17 at 1550°C and decreasing probably to about 0.12 at 1500°C. Using these activity data and equation (6.9) the deoxidation equilibria are calculated for Al/Si/Mn; these are compared in Fig. 6.18 with the residual ppm O for the Si/Mn deoxidation at the same concentrations of Mn and Si.

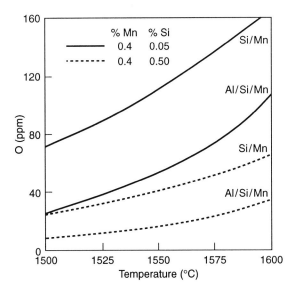

Fig. 6.18   Deoxidation equilibria with Si/Mn compared with Al/Si/Mn for the deoxidation product saturated with $Al_2O_3$.

### 6.8.4 DEOXIDATION WITH Al

Numerous laboratory experiments have been made on the aluminium deoxidation of liquid steel using the EMF technique for measuring the oxygen activity in the melt. The equilibrium constants obtained from independent experimental studies agree within about a factor of two. An average value for the quilibrium constant is given below.

$$Al_2O_3(s) = 2[Al] + 3[O] \tag{6.26}$$

$$K = \frac{[\%Al]^2[ppm\ O \times f_O]^3}{a_{Al_2O_3}} \tag{6.26a}$$

$$\log K = -\frac{62{,}680}{T} + 31.85 \tag{6.26b}$$

The alumina activity is with respect to pure solid $Al_2O_3$. The effect of aluminium on the activity coefficient of oxygen dissolved in liquid steel is given by $\log f_O = -3.9 \times [\%Al]$. At low concentrations of aluminium, $f_{Al} \approx 1.0$.

It should be noted that in the commercial oxygen sensors the electrolyte tip is MgO-stabilised zirconia. At low oxygen potentials as with aluminium deoxidation, there is some electronic conduction in the MgO-stabilised zirconia which gives an emf reading that is somewhat higher than $Y_2O_3$ or $ThO_2$ stabilised zirconia where the electronic conduction is negligibly small. In other words, for a given concentration of Al in the steel the commercial oxygen sensor, without correction for partial electronic conduction, registers an oxygen activity that is higher than the true equilibrium value. To be consistent with the commercial oxygen sensor readings, the following apparent equilibrium constant may be used for reaction (6.26) for pure $Al_2O_3$ as the reaction product.

$$\log K_a = -\frac{62{,}680}{T} + 32.54 \tag{6.27}$$

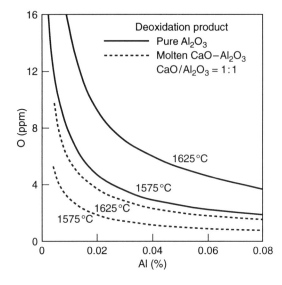

Fig. 6.19   Deoxidation with aluminium in equilibrium with $Al_2O_3$ or molten calcium aluminate with $CaO/Al_2O_3 = 1{:}1$.

When the Al-killed steel is treated with Ca–Si the alumina inclusions are converted to molten calcium aluminate. For the ratio $\%CaO/Al_2O_3 = 1:1$, the activity of $Al_2O_3$ is 0.064 with respect to pure $Al_2O_3$ at temperatures in the range 1500–1700°C. The apparent equilibrium relations, consistent with the readings of commercial oxygen sensors, are shown in Fig. 6.19 for the deoxidation products: pure $Al_2O_3$ and molten calcium aluminate with $\%CaO/\%Al_2O_3 = 1:1$.

## REFERENCES

1. P. BREMER, *Stahl u. Eisen*, 1951, **71**, 575.
2. R.V. PATHY and R.G. WARD, *J. Iron and Steel Inst.*, 1964, **202**, 995.
3. E. AUKRUST, P.J. KOROS and H.W. MEYER, *J. Met.*, 1966, **18**, 433.
4. F.D. RICHARDSON and W.E. DENNIS, *J. Iron and Steel Inst.*, 1953, **175**, 257, 264.
5. E.T. TURKDOGAN, *Physicochemical properties of molten slags and glasses*, The Metals Society (now The Institute of Materials), London, 1983.
6. H. KNUPPEL and F. OETERS, *Stahl u. Eisen*, 1961, **81**, 1437.
7. H. SUITO, R. INOUE and M. TAKADA, *Trans. Iron and Steel Inst, Japan*, 1981, **21**, 250; 1982, **22**, 869.
8. R. NAGABAYASHI, M. HINO and S. BAN-YA, *Tetsu-to-Hagané*, 1988, **74**, 1577.
9. J.C. WRAMPELMEYER, S. DIMITROV and D. JANKE, *Steel Research*, 1989, **60**, 539.
10. T.B. WINKLER and J. CHIPMAN, *Trans. AIME*, 1946, **167**, 111.
11. C.H. HERTY, *Trans. AIME*, 1926, **73**, 1107.
12. H.O. VON SAMSON-HIMMELSTJERNA, *Arch Eisenhüttenwes*, 1932, **6**, 471.
13. W. BISCHOF and E. MAURER, *Arch. Eisenshüttenwes*, 1932, **6**, 415.
14. E. MAURER and W. BISCHOF, *J. Iron and Steel Inst.*, 1935, **132**, 13.
15. G. TROMEL and K. SCHWERDTFEGER, *Arch. Eisenhüttenwes*, 1963, **34**, 101.
16. S. BAN-YA, T. WATANABE and R. NAGABAYASGI, in *Advances in the science of iron and steelmaking*, p. 79. Japan/USA Seminar, Kyoto, May 16–28, 1983.
17. E.T. TURKDOGAN, *Ironmaking and Steelmaking.*, 1988, **15**, 311.
18. B. OZTURK and E.T. TURKDOGAN, *Metal Science*, 1984, **18**, 299.
19. R.A. SHARMA and F.D. RICHARDSON, *J. Iron and Steel Inst.*, 1961, **198**, 386.
20. G.J.W. KOR and F.D. RICHARDSON, *J. Iron and Steel Inst.*, 1968, **206**, 700.
21. F. KORBER and W. OELSEN, *Mitt. Kaiser-Wilhelm Inst. Eisenforsch.*, 1935, **17**, 231; 1933, **15**, 271.
22. B.K.D.P. RAO and D.R. GASKELL, *Metall. Trans. B.*, 1981, 12B, 311.
23. K.P. ABRAHAM, M.W. DAVIES and F.D. RICHARDSON, *J. Iron and Steel Inst.*, 1960, **196**, 82.

CHAPTER 7

# Pretreatment of Blast Furnace Iron

The impetus to hot metal refining developed by the Japanese steel industry was, in part, to accomplish the objective of improving cost effectiveness in steelmaking by reducing the volume of waste product slag to be disposed of.

The hot metal refining, developed primarily by the steel industry in Japan, involves two or three processing steps. The first step is desiliconisation to residual Si < 0.15 percent. After deslagging, the hot metal is desulphurised and dephosphorised by injecting with nitrogen plus oxygen a mixture of sinter fines, burnt lime, calcium fluoride and some calcium chloride. In some practices, sodium carbonate alone is injected. There are many variations in the method of hot metal refining practices in the Japanese steel industry. As examples, the flow diagrams of three different refining processes are shown in Figs. 7.1, 7.2, 7.3. For a comprehensive review of these developments in the late 1970s, reference may be made to the papers by Fuwa[1] and Ishihara.[2]

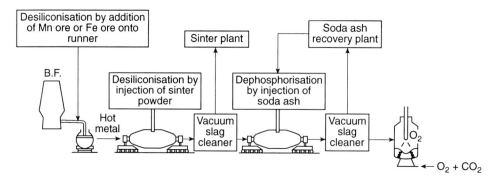

Fig. 7.1  Hot metal refining with $Na_2CO_3$ in torpedo car – Kashima Works of Sumitomo Metal Industries.

## 7.1 Desiliconisation

To improve steelmaking in open hearth furnaces, attempts were made in the 1940s in the European and North American steel plants to lower the silicon content of hot metal from more than 1% Si to 0.4–0.6% Si. This was done simply by adding dried mill scale either onto the blast furnace runner or into the hot metal transfer ladle.

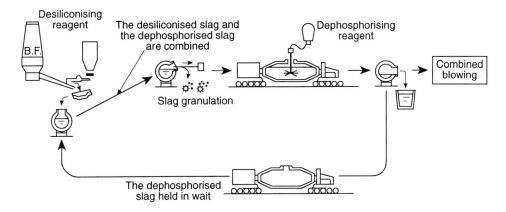

Fig. 7.2   Hot metal refining with lime–scale–salt in torpedo car – Kimitsu Works of Nippon Steel Corporation.

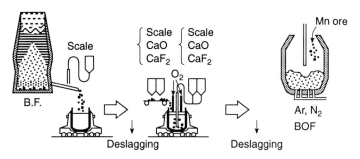

Fig. 7.3   Hot metal refining with lime-scale–salt in transfer ladle – Keihin Works of NKK Corporation.

Table 7.1   Desiliconisation Processes

| Process | Reagent usage for $\Delta Si = 0.4\%$ | Efficiency % |
|---|---|---|
| Mill scale and lime added to BF runner | 15–25 kg t⁻¹ mill scale 8–10 kg t⁻¹ lime | 80–90 |
| Mill scale added to transfer ladle | 32 kg t⁻¹ | 60–80 |
| Mill scale injection into torpedo car | 15–25 kg t⁻¹ for 0.25% $\Delta Si$ | 40–60 |
| Oxygen blowing into torpedo car | 0.13 Nm³ min⁻¹t⁻¹ | 35–40 |
| Deep oxygen injection into transfer ladle | 0.3 Nm³ min⁻¹t⁻¹ | 45–55 |

In more sophisticated present practices, the hot metal is desiliconised from 0.4–0.6% to less than 0.15%, to facilitate the subsequent stage of refining for phosphorus and sulphur removal. Various methods of desiliconisation are listed in Table 7.1.

Depending on the operating conditions, there is a temperature gain or loss of about ± 20°C when mill scale or ore sinter fines are used for desiliconisation. Although the oxidation of silicon with iron oxide is an exothermic reaction, a temperature loss in hot metal may occur because of slag removal, using moist reagents, extra fluxes or carrier gases. At low levels of initial silicon content some decarburisation does occur during desiliconisation which also contributes to a drop in melt temperature. With oxygen lancing the hot metal temperature will increase by between 120°C and 150°C.

The slags produced during desiliconisation contain primarily $SiO_2$, FeO, MnO and CaO. Since the phosphorus and sulphur contents of these slags are low they can be recycled to the sinter plants.

## 7.2 Dephosphorisation & Desulphurisation

After desiliconisation and slag removal, the hot metal is dephosphorised and desulphurised with the injection of an oxidising basic flux mixture using nitrogen or air as a carrier gas and accompanied with top or bottom blowing of oxygen in some processes. Examples of processes and fluxes used are summarised in Table 7.2.

**Table 7.2**   Dephosphorisation & Desulphurisation Processes

| Process | Vessel | Flux mixture, % | kgt⁻¹ | % P Before | % P After | % S Before | % S After |
|---|---|---|---|---|---|---|---|
| Injection with $N_2$ | Torpedo car | 35 CaO, 55 mill scale 5 CaF$_2$, 5 CaCl$_2$ | 52 | 0.10 | 0.015 | 0.025 | 0.005 |
| Injection with $N_2$ $O_2$ top blowing | Torpedo car | Na$_2$CO$_3$ | 15 20 | 0.09 0.11 | 0.011 0.020 | 0.040 0.060 | <0.01 <0.01 |
| Na$_2$CO$_3$ top addition $O_2$ top blowing $N_2$ bubbling | Ladle | Na$_2$CO$_3$ | 20 | 0.10 | <0.010 | 0.030 | <0.005 |
| Injection with $N_2$ | Ladle | 30 CaO, 62 sinter fines 4 CaF$_2$ 4 CaCl$_2$ | 45 | 0.12 | 0.010 | 0.025 | <0.01 |
| Injection with $N_2$ $O_2$ top blowing | Ladle | 38 CaO, 42 sinter fines 20 CaF$_2$ | 40 | 0.10 | 0.010 | 0.040 | 0.02 |
| $O_2$ bottom blowing | Q–BOP | 39 CaO, 55 sinter fines 6 CaF$_2$ | 51 | 0.14 | 0.010 | 0.020 | 0.01 |

As is seen from the experimental and plant data in Fig. 7.4 reported by Marukawa[3], high phosphorus distribution ratios between slag and metal are obtained at a lower temperature treatment of hot metal with soda ash. A major drawback to using sodium carbonate-based fluxes is in the evolution of alkali fumes from the reduction of carbonate by carbon in iron

$$Na_2CO_3 \ (l) + 2[C] \rightarrow 2 \ Na(g) + 3 \ CO \ (g) \qquad (7.1)$$

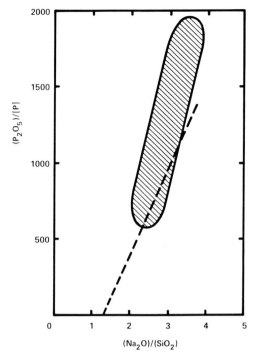

Fig. 7.4 Phosphorus distribution between sodium silico-phosphate slags and graphite-saturated liquid iron at 1300 to 1350°C: – – – – crucible experiments; soda ash injection in the ladle (250 t). From Ref. 3.

The higher the ratio $Na_2O/SiO_2$ in the slag and higher the melt temperature, the greater would be the generation of alkali fumes.

Because of the pronounced effect of temperature on the dephosphorisation of iron, the refining of hot metal is done at low melt temperatures. The plant data in Fig. 7.5 from Keihin Works of NKK Corp. as reported by Sinde *et al.*[4], demonstrate to what extent hot metal in the ladle can be dephosphorised. The melt temperature is controlled by adjusting the ratio $O_2(gas)/\{O_2(gas) + O_2 \text{ (scale)}\}$ using the relation in Fig. 7.6 obtained from plant data. The slag/metal phosphorus distribution ratio at the end of refining increases with an increasing slag basicity up to about $CaO/SiO_2 = 4.5$ as depicted in Fig. 7.7. At slag basicities > 4.5 the slag contains undissolved lime which hinders effective slag-metal mixing, hence lowering the extent of metal dephosphorisation.

The plant data in Fig. 7.8 from Wakayama Works of Sumitomo Metal Industries, Ltd.,[5] show changes in hot metal composition with the treatment time. For an average refining time of about 30 minutes, the carbon content of the hot metal decreases by 0.2 to 0.3 percent.

High slag/metal phosphorus distribution ratios obtained in hot metal refining are much greater than would be predicted from the phosphate

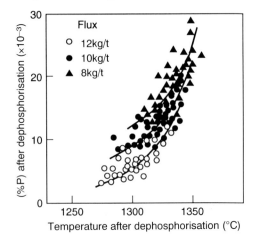

Fig. 7.5 Influence of temperature and flux consumption on [%P] after dephosphorisation.

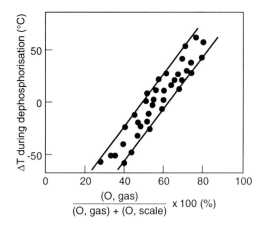

Fig. 7.6 Relation between the gaseous oxygen ratio and ΔT during dephosphorisation.

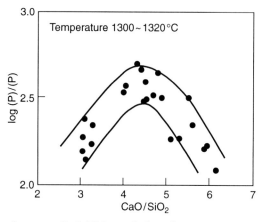

Fig. 7.7 Relation between $CaO/SiO_2$ and phosphorus partition ratio.

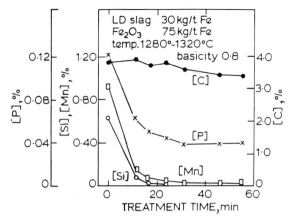

Fig. 7.8 Changes in hot metal composition during dephosphorisation treatment from data reported by Nashiwa *et al.*[5]

capacities of the slags, on the assumption of carbon-oxygen equilibrium in the melt. An extensive degree of dephosphorisation achieved in hot metal refining, is brought about by the non equilibrium state of high oxygen activity in the melt during the flux injection with nitrogen plus oxygen. Takeuchi *et al.*[6] made oxygen sensor measurements at different positions in the melt during the flux injection with a 75% $O_2$ + 25% $N_2$ mixture. They found that the measured oxygen activity in the melt was about 1000 times greater near the lance tip and about 100 times greater near the slag layer in comparison to the C–O equilibrium.

## 7.3 DESULPHURISATION

In most steel plants outside Japan, hot metal refining is confined to desulphurisation in the transfer ladle with various injected materials such as lime plus spar, lime plus magnesium, calcium carbide plus magnesium or calcium carbide plus limestone.

Plant data from various steel works are plotted in Fig. 7.9 for three types of reagents injected at the rates of 30 to 40 kg min$^{-1}$. As shown in Fig. 7.10 the quantity of Mg to be injected together with CaO or $CaC_2$ is adjusted in accord with the initial sulphur content of the hot metal. Desulphurisation occurs primarily by the following reactions

$$Mg(g) + [S] \rightarrow MgS(s)$$
$$CaC_2 \rightarrow Ca(g) + 2[C]$$
$$Ca(g) + [S] \rightarrow CaS(s)$$

Most of the desulphurisation is done by magnesium. For the removal of 0.04% S in the desulphurisation of hot metal, with about 70% efficiency of magnesium usage, the amount of the reagent to be injected would be 1.45

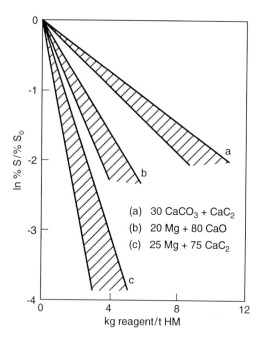

Fig. 7.9   Hot metal desulphurisation with lance injection in transfer ladle.

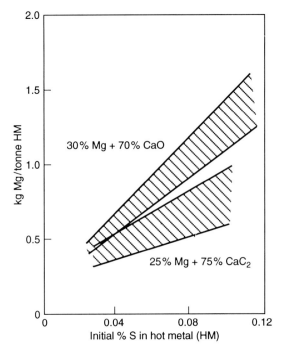

Fig. 7.10   Plant data within shaded area showing magnesium usage in hot metal desulphurisation with Mg+CaO and Mg+CaC$_2$ injection in the transfer ladle at the rate of 30 to 40 kg min$^{-1}$.

kg t$^{-1}$ HM for the 30 Mg + 70 CaO mixture. For this quantity of solid injection with N$_2$ flowing at the rate of 35 Nl per kg solids, the heat absorbed from the melt would be about 4310 kJ t$^{-1}$ HM. On the other hand, the heat generated by the reaction Mg(g) + [S] → MgS would be about 5060 kJ t$^{-1}$ HM for 0.04% ΔS which compensates for the thermal energy absorbed in heating the injected material to the bath temperature.

The solubility of MgS in hot metal decreases with a decreasing bath temperature thus

$$\log [\%Mg][\%S] = -\frac{17,026}{T} + 5.15$$

A lower residual dissolved sulphur will be achieved at a lower refining temperature. Depending on the operating conditions the residual dissolved sulphur in the treated hot metal is in the range 0.002 to 0.005% S. The sulphur-rich slag in the transfer ladle is skimmed off as much as operating conditions will permit, prior to charging the hot metal to the BOF vessel.

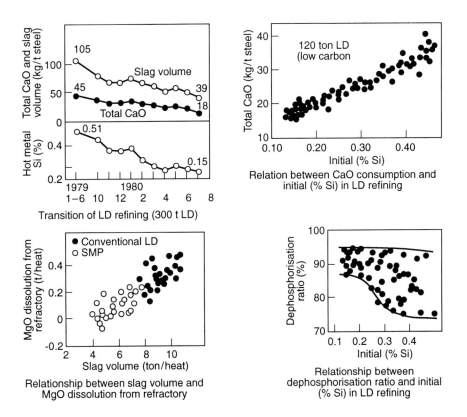

Fig. 7.11   Selected diagrams from a paper by Itoh *et al.*[8] on the development of slag minimum refining process (SMP) by Nippon Steel Corporation.

During desulphurisation the nitrogen content of hot metal is reduced from 60–70 ppm to 30–35 ppm which reflects on lower levels of nitrogen in BOF heats at tap. For example, for 30 to 35 ppm N in desulphurised hot metal the BOF steel at tap would contain < 20 ppm N, as observed at the steel works of Thyssen Stahl AG, Duisburg.[7]

### 7.4 BENEFITS OF HOT METAL REFINING

It is seen from the plant data of Muroran Works of Nippon Steel Corporation in Fig. 7.11 reported by Itoh *et al.*[8], that by reducing the silicon content of hot metal to <0.2%, a substantial saving was realised in lime consumption and the extent of dephosphorisation became more stable. As would be expected, a decrease in slag volume also lowers the extent of refractory wear and increases the steel yield. The minimum slag practice decreases scrap melting in BOF which proves to be an advantage for easier control of the steelmaking operation.

The overall economics of the minimum slag practice in BOF steelmaking depends a great deal on the availability and the relative cost of raw materials, steel scrap and hot metal. At present the minimum slag practice, with 95 percent or more refined hot metal in the BOF charge, is neither practical nor cost effective in most steel industries outside Japan.

### REFERENCES

1. T. FUWA, *Iron and Steelmaker*, 1981, **8**(6), 18–25; **8**(7), 25.
2. S. ISHIHARA, *Trans. Iron Steel Inst. Japan*, 1985, **25**, 537.
3. K. MARUKAWA, *Development of the Preliminary Treatment Process of Molten Pig Iron with Sodium Carbonate Flux*, Report, 1982, Kashima Works of Sumitomo Metal Industries, Ltd.
4. T. SINDE *et al.*, *Tetsu-to-Hagané*, 1987, **73**(12), S862.
5. H. NASHIWA, T. ADACHI, T. OKAZAKI and K. IEDA, *Ironmaking & Steelmaking*, 1981, **8**(1), 29.
6. S. TAKEUCHI, M. OZAWA, T. NOZAKI, T. EMI and M. OHTANI, in 'Development of Hot Metal Preparation for Oxygen Steelmaking', 13–1; McMaster University, Hamilton, Ontario, 1983.
7. R.W. SIMON, W. FLORIN, R. HAMMER and E. HÖFFKEN, *Steelmaking Conference Proceedings*, 1989, **72**, 313.
8. Y. ITOH, S. SATOH and Y. KAWAUCHI, *Trans. Iron Steel Inst. Japan*, 1983, **23**, 256.

# CHAPTER 8
# Oxygen Steelmaking

## 8.1 INTRODUCTION

The first commercial operation of steelmaking with oxygen top blowing in the converter was in the early 1950s at Linz and Donawitz (Austria). This manner of steelmaking became known as Linz–Donawitz or LD process. For many years now, most of the steel has been made by top oxygen blowing for which different names are given. For example, in European steel plants the process is still called LD; in the UK, BOS (basic oxygen steelmaking); in the Far East and America, BOF (basic oxygen furnace), with the exception of U.S. Steel where it is called BOP (basic oxygen process).

In the early 1970s, a bottom-blown oxygen steelmaking process was developed in Canada and Germany. This process, known as OBM in Europe and Q–BOP elsewhere, was in full size commercial operation by the mid 1970s in U.S. Steel plants followed by several plants in Europe and Japan. The tuyeres, mounted in a removable bottom, consist of a central pipe for blowing oxygen together with burnt lime, and an annular gap around the central pipe for the passage of gaseous hydrocarbon, e.g. propane or natural gas ($CH_4$). Upon contact with liquid steel the hydrocarbon dissociates to C and $H_2$ with the absorption of heat. This endothermic reaction suppresses overheating of the tuyere tip by the exothermic reaction of oxygen with liquid steel.

Further developments in oxygen steelmaking led to the present practices of various types of top and bottom blowing known as combined blowing, as illustrated schematically in Fig. 8.1 for BOF and Q–BOP processes. There is also post combustion of CO in the upper part of the vessel to generate additional heat for steelmaking.

The presentation in this chapter is on the fundamentals of steelmaking reactions and assessment of the state of slag–metal reactions at the end of oxygen blowing, i.e. at first turndown of the vessel for sample taking. For a description of steelmaking facilities and plant layout, reference may be made to the U.S. Steel publication *The Making, Shaping and Treating of Steel*, 1985 edition.

## 8.2 FURNACE CHARGE

Depending on the local operating conditions, availability of scrap, blast furnace iron (hot metal) and the extent of hot metal pretreatment, 75 to 95

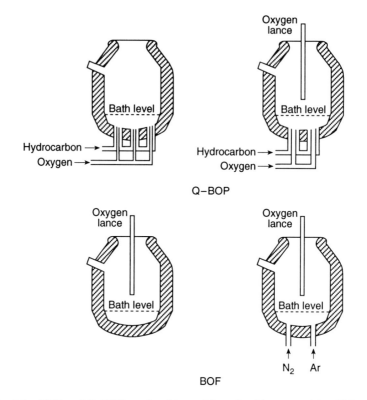

Fig. 8.1   BOF and Q–BOP steelmaking with and without combined blowing.

percent of the metallic charge to the BOF and Q–BOP vessel is hot metal and the remainder steel scrap. Generally speaking the desulphurised hot metal has the following composition in mass percent:

$$5 \text{ C, } 0.4 - 0.7 \text{ Mn, } 0.4 - 1.0 \text{ Si,}$$
$$0.05 - 0.10 \text{ P and } 0.002 - 0.005 \text{ S}$$

Types of scrap used are generally those produced in the steel mill: sheet scrap, slab ends, bloom ends, cold iron or broken moulds, pit scrap, bi-metallic cans and hot crop. Then there is the purchased scrap such as scrapped auto bodies. It is important that scrap composition be considered when calculating a scrap charge for any particular heat. Residuals in the scrap such as tin, copper, nickel, molybdenum and tungsten are not ox-idised in the steelmaking process and could be undesirable in the final product. Scrap density is also very important. Scrap is classified as 'light' or 'heavy'. Most steelmaking shops use a 1:1 ratio of light to heavy scrap in the furnace charge. It is advantageous to charge the light scrap into the vessel first to minimise wear and tear of the furnace refractory lining.

Some alloy additions are also included in the furnace charge to meet steel specifications: (i) copper, commercial grade or 22% Cu–Fe alloy, (ii) nickel as nickel oxide or 48% Ni–Fe alloy, (iii) molybdenum as moly oxide and (iv) chromium as an iron alloy. When the silicon content of hot metal is low, some silicon carbide and/or ferrosilicon are included in the furnace charge as additional fuel for steelmaking.

Soon after starting oxygen blowing, burnt lime and burnt dolomite are charged into the vessel as fluxes. In some shops a small addition of fluor-spar ($CaF_2$) is made when the aimed carbon content of the steel at turndown is more than 0.10 percent. In some practices, limestone, dolomite and/or iron ore are used as a coolant when needed. In the Q-BOP steelmaking all the lime needed for the process is bottom blown together with oxygen.

### 8.2.1 STATIC CHARGE CONTROL

Computer-aided charge-control calculations are made for every heat. About 80 percent of the charge-control model is based on the heat and material balance, the remainder being based on empirical relations which vary from one melt shop to another. Since every steelmaking shop has its own formulation of the charge-control model, only general aspects of this subject will be discussed here in a simplified form.

#### 8.2.1a Material balance

Steelmaking is an oxidation process to remove the oxidisable elements from hot metal and scrap to the furnace slag as the steel is decarburised with oxygen blowing.

$$[C] + \tfrac{1}{2}O_2 \quad \rightarrow \quad CO \text{ (gas)}$$

$$\left. \begin{array}{l} [Si] + O_2 \quad \rightarrow \quad SiO_2 \\[4pt] [P] + \tfrac{5}{4}O_2 \quad \rightarrow \quad \tfrac{1}{2}P_2O_5 \\[4pt] [Mn] + \tfrac{1}{2}O_2 \quad \rightarrow \quad MnO \\[4pt] Fe + \tfrac{1}{2}O_2 \quad \rightarrow \quad FeO \end{array} \right\} \quad \begin{array}{l} \text{Fluxed with} \\ \text{lime (CaO) forming} \\ \text{molten slag} \end{array}$$

At the end of oxygen blowing the slag basicity (as $V = \%CaO/\%SiO_2$) is in the range 2.5 to 4.0, with iron oxide contents of 10 to 35% FeO depending on the aimed carbon, sulphur and phosphorus contents of the steel at tap. From trial and error calculations the following formulation has emerged for the quantities of burnt lime and burnt dolomite to be charged, depending on the total silicon content of the furnace charge.

$$\text{kg burnt lime/t metallic charge} = 19 \times V \times \%Si \text{ (in furnace charge)} \tag{8.1}$$

$$\frac{\text{kg burnt lime}}{\text{kg burnt dolomite}} = (2 + 0.3 \times V) \tag{8.2}$$

Calcined (burnt) dolomite, known as 'doloma' or 'dolo', is always included in the furnace charge to minimise slag attack on the MgO–C brick lining of the vessel. Burnt dolomite contains about 56% CaO and 41% MgO the remainder being impurities. Burnt lime contains about 96% CaO, 1% MgO, 1% $SiO_2$ and other minor impurities with a 1.5% loss of ignition.

In making an approximate calculation of the material balance given here as an example, certain average values are used with regard to the composition of slags at tap as given below.

| $V = \%CaO / \%SiO_2$ | %FeO(total) | %MgO |
|---|---|---|
| 2.5 | | 6.7 |
| 3.0 | $(16.6 \times V - 30)$ | 5.4 |
| 3.5 | | 4.7 |
| 4.0 | | 4.3 |

In steelmaking slags the sum of the four primary oxides CaO + MgO + FeO + $SiO_2$ is in the range 88 to 92%. In calculating the MgO solubilities for the indicated $V$ ratios, using the data in Fig. 5.12 for the quaternary system, the values are adjusted to an average value of 90% for the sum of these four oxides.

The slag mass per tonne of steel is estimated from the quantities CaO, MgO, Si (as $SiO_2$) in the furnace charge and $X\%$ FeO in the tap slag for the aimed slag basicity V as given below.

$$\text{Slag mass, kg t}^{-1} \text{ steel} = \frac{CaO + MgO + SiO_2}{90 - (16.6 \times V - 30)} \tag{8.3}$$

Two examples of calculations of the furnace charge are given using the foregoing formulations.

Example *A* for 0.4% Si in the furnace charge with an aimed slag basicity $V$ = 2.5:

| | | |
|---|---|---|
| Burnt lime : | $19 \times 2.5 \times 0.4$ | = 19 kg t$^{-1}$ |
| Doloma    : | 19/2.75 | = 6.91 kg t$^{-1}$ |
| CaO       : | $19 \times 0.96 + 6.91 \times 0.57$ | = 22.18 kg t$^{-1}$ |
| MgO       : | $6.91 \times 0.4 + 0.19$ | = 2.95 kg t$^{-1}$ |
| $SiO_2$     : | $4 \times 60/28 + 0.19$ | = 8.76 kg t$^{-1}$ |
| Slag mass = | $\dfrac{2218 + 295 + 876}{90 - 11.5}$ | = 43.17 kg t$^{-1}$ |

Estimated slag composition at tap:
51.38% CaO, 6.83% MgO, 11.5% FeO and 20.29% $SiO_2$; $V$ = 2.53.

Example *B* for 1% Si in the furnace charge with the aimed slag basicity $V$ = 4.0:

| Burnt lime : | $19 \times 4.0 \times 1$ | $= 76.0$ kg t$^{-1}$ |
|---|---|---|
| Doloma : | $76/3.2$ | $= 23.75$ kg t$^{-1}$ |
| CaO : | $76 \times 0.96 + 23.75 \times 0.57$ | $= 86.50$ kg t$^{-1}$ |
| MgO : | $23.75 \times 0.4 + 0.76$ | $= 10.26$ kg t$^{-1}$ |
| SiO$_2$ : | $10 \times 60/28 + 0.76$ | $= 22.19$ kg t$^{-1}$ |
| Slag mass $=$ | $\dfrac{8650 + 1026 + 2219}{90 - 36.4}$ | $= 221.92$ kg t$^{-1}$ |

Estimated slag composition at tap:
38.98% CaO, 4.62% MgO, 36.4% FeO and 10% SiO$_2$; $V = 3.90$

In the detailed formulation of material balance for the furnace charge, several additional terms are incorporated in the charge control system, as for example: aimed turndown temperature, carbon, sulphur and phosphorus contents in the steel, also the slag volume required to achieve the aimed steel composition at tap. These additional terms for the charge control model are formulated using reliable equilibrium data on slag–metal reactions, with plant analytical data logs on slag and metal samples taken from the vessel at tap.

*8.2.1b Heat balance*

For a 220 tonne metallic furnace charge containing 4% C, 0.6% Si, 0.5% Mn and 0.06% P to produce steel containing at tap 0.03% C, <0.005% Si, 0.2% Mn and 0.01% P, the thermal energy generated by oxidation will be as follows.

| [C] | + | ½O$_2$ | $\rightarrow$ | CO | $- 83.4 \times 10^6$ kJ |
|---|---|---|---|---|---|
| [Si] | + | O$_2$ | $\rightarrow$ | SiO$_2$ | $- 35.4 \times 10^6$ kJ |
| [P] | + | ⁵⁄₄O$_2$ | $\rightarrow$ | ½P$_2$O$_5$ | $- 2.2 \times 10^6$ kJ |
| [Mn] | + | ½O$_2$ | $\rightarrow$ | MnO | $- 4.0 \times 10^6$ kJ |
| Fe | + | ½O$_2$ | $\rightarrow$ | FeO | $- 17.1 \times 10^6$ kJ |
| | | Total heat generated $=$ | | | $-142.1 \times 10^6$ kJ |

The heat generation due to the oxidation of iron is for a 22 t furnace slag containing 20% FeO (total) which corresponds to the oxidation of 3421 kg of iron.

The approximate quantities of heat requirement to raise the bath temperature to 1650°C are given below.

| Amount and Temperature of Charge Material | Heat Required |
|---|---|
| 175 t  *HM*, 1340°C | $38.21 \times 10^6$ kJ |
| 45 t  steel scrap, 25°C | $60.77 \times 10^6$ kJ |
| 7.3 t  lime  $\Big\}$ 25°C<br>3.0 t  doloma | $14.31 \times 10^6$ kJ |
| Total $=$ | $113.29 \times 10^6$ kJ |

The difference of about $29 \times 10^6$ kJ between the quantities of heat generated and heat absorbed is a measure of the heat losses, primarily by the furnace off gases (CO) and in part by heat conduction through the vessel lining and outer shell into the melt shop environment.

Examples of heat balances are given below for reactions of various furnace charges in the vessel with respect to the furnace charges being at room temperature and the steel bath at 1600°C.

| Reaction | $\Delta H$, kJ kg$^{-1}$ furnace charge |
|---|---|
| $Fe_2O_3 + 3[C] \rightarrow 2[Fe] + 3CO$ | 4607 |
| $NiO + [C] \rightarrow [Ni] + CO$ | 3362 |
| $MoO_2 + 2[C] \rightarrow [Mo] + 2CO$ | 4527 |
| $CaCO_3 + [C] \rightarrow CaO + 2CO$ | 5614 |
| $Cu \rightarrow [Cu]$ | 924 |
| $Fe-75\% \, Si + O_2 \rightarrow [Fe] + SiO_2$ | −20,844 |
| $SiC + \frac{3}{2}O_2 \rightarrow SiO_2 + CO$ | −19,420 |

To generate additional heat in the vessel for minimising skull buildup at the converter mouth, some of the carbon monoxide generated in the steel decarburisation is combusted with oxygen in the upper part of the vessel. For $O_2$ at 25°C, and CO and $CO_2$ at 1600°C, the heat generated would be

$$CO + \frac{1}{2}O_2 \rightarrow CO_2, \, \Delta H = -22,420 \text{ kJ Nm}^{-3} \, O_2$$

In the 220 tonne heats, about 160 Nm³$O_2$ is used for post combustion during the oxygen blowing time of about 22 minutes in BOF and 16 to 18 minutes in Q–BOP steelmaking.

*8.2.1c Reliability of static charge-control system*

The time sequence of charge-control calculations and materials handling is shown in Fig. 8.2 in a simplified form. The preliminary calculation for heat B starts at the time of charging the vessel for heat A. How good of an estimate can be made of the temperature and composition of the steel at the end of oxygen blowing (on the basis of the static charge control programme) depends on (i) accuracy of the charge control model, (ii) accuracy of inputs to the computer system, (iii) consistency of steelmaking practices and quality of materials used, (iv) reliability of the computer system and its use, and (v) reliability of measuring and interfacing devices.

**8.3 DYNAMIC CONTROL OF OXYGEN BLOWING**

It was in the 1970s that the sublance technology was developed in the Japanese steel industry for dynamic control of oxygen steelmaking. The

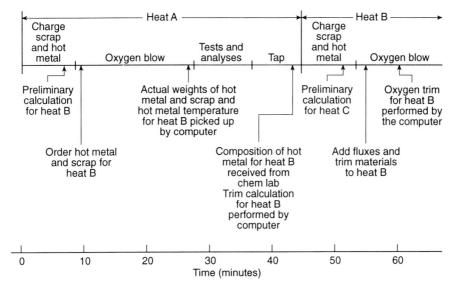

Fig. 8.2  Time sequence of charge-control calculations and materials handling.

dynamic control system involves the use of a sublance for sampling and analysing waste gas for CO and $CO_2$ together with the computer model for charge control. A water-cooled sublance is lowered into the steel bath to measure the bath temperature, the carbon content of steel and to take a metal sample for chemical analysis. This sampling is done usually 2 or 3 minutes before the scheduled end of the blow. With the data collected, including the waste-gas analysis, calculations are made by the process-control computer to determine the corrections to the blow that are necessary to achieve the desired end point temperature and carbon content of the steel. Use of the sublance with a good static charge control model and dynamic process control system will ascertain first turndown hit rates of over 90 percent.

### 8.3.1 AUTOMATED CONTROL OF OXYGEN BLOWING

Use of the sublance in controlling oxygen blowing varies from one steel mill to another. A brief description is given here of a fully automatic oxygen blowing practice that was developed at Kawasaki Steel Corporation, reported in a paper by Iida *et al.*[1]

As shown in Fig. 8.3, the automatic control system for blowing consists of four subsystems.

*A: Subsystem for static control*

The raw material blending and the amounts of oxygen, coolant and fluxes necessary for blowing, are calculated using static control models based on the material balance and heat balance.

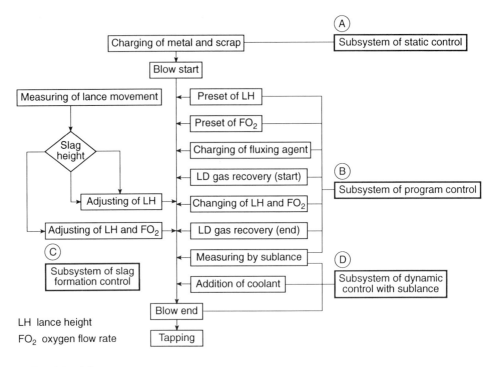

Fig. 8.3   Schematic representation of the system and function of fully automatic BOF steelmaking. From Ref. 1.

*B: Subsystem for programme control*

The lance height and oxygen blowing rate are automatically controlled and the fluxes automatically charged according to the computer programme selected before starting the blow. This subsystem also controls the operation of the waste gas treatment system, the gas recovery operation and start of the sublance operation.

*C: Subsystem for slag formation*

This subsystem continuously monitors slag formation based on the measured values of acceleration of lance vibration, which controls the lance height and oxygen flow rate, so that the slag foam height is maintained at the desired level.

*D: Subsystem for dynamic control by sublance*

The steel temperature and carbon content are measured by the sublance during blowing and the required amounts of oxygen and coolant are calculated using the dynamic control model, whereby charging of the coolant and terminating oxygen blowing are automatically carried out.

It should be noted that a highly automated oxygen steelmaking is practical and cost effective, only when more than 90% of the furnace charge is pre-treated hot metal with Si<0.2%, P<0.01% and S<0.01%, as accomplished in many of the Japanese steel works.

## 8.4 STEELMAKING SLAGS

### 8.4.1 SLAG FORMATION IN BOF STEELMAKING

In BOF steelmaking the metal and slag compositions change during oxygen blowing as shown in Fig. 8.4. These data, reported by van Hoorn *et al.*[2], were obtained from a series of trial heats made in a 300 t capacity BOF at Hoogovens IJmuiden BV. In the early stages of the blow most of the silicon is oxidised forming a slag of low basicity. Humps on the manganese and phosphorus curves are characteristic of all pneumatic steelmaking processes caused by changes in the melt temperature and slag composition.

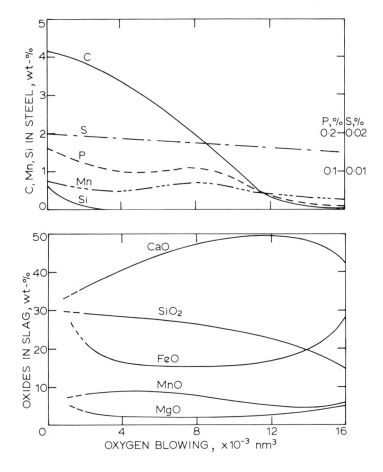

Fig. 8.4   Changes in metal and slag compositions during steelmaking in the BOF for about a 300 t melt, using data of van Hoorn *et al.*[2].

Changes in slag composition during oxygen blowing depend on the BOF practice which varies from one steelworks to another. Changes in slag composition are shown in Fig. 8.5a for two different practices; in this graphic representation the slag composition is recalculated to give CaO + MgO + FeO + SiO$_2$ = 100%. The curve I is that reported by van Hoorn *et al.* for the practice at Hoogovens IJmuiden BV. This practice is claimed to lead to low metal dispersion in the slag and minimum slopping of the bath. It is also considered to be good for a fast rate of decarburisation and most suitable for low sulphur and phosphorus in the furnace charge. The overoxidised-slag practice of Mannesmann is represented by curve II as reported by Bardenheurer *et al.*[3] With this practice a liquid slag of high basicity is obtained more readily early in the blow, resulting in a low magnesia pickup by the slag and a faster rate of removal of sulphur and phosphorus. With this practice however there is a greater tendency to slopping. Two other practices are shown in Fig. 8.5b. According to Nilles *et al.*[4] a path AA for slag formation gives the best refining conditions,

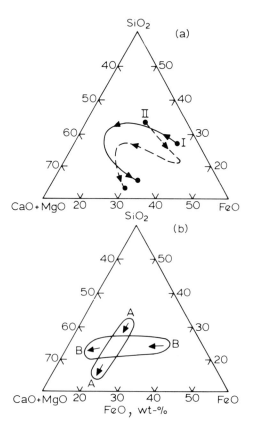

Fig. 8.5   Changes in composition of slag during oxygen blowing for various BOF practices: (a) I, van Hoorn *et al.*[2] and II, Bardenheurer *et al.*[3] (b) AA, Nilles *et al.*[4] and BB, Baker[5].

particularly for decarburisation. The path BB was considered by Baker[5] to be better for phosphorus and sulphur removal.

In the early stages of the blow the slag basicity is low; therefore the solubility of MgO is high. As the basicity increases with the progress of oxygen blowing, the solubility decreases resulting in rejection of MgO from the molten slag. However, an increase in slag basicity during the blow is accompanied by an increase in slag mass which counteracts MgO rejection; in fact the total amount of MgO taken up by the slag increases during the later stages of the blow.

### 8.4.2 SLAG FORMATION IN Q-BOP STEELMAKING

In Q-BOP steelmaking the oxygen and lime powder are blown through a series of tuyeres located on the bottom of the converter. The mode of slag formation during the blow in Q–BOP will be somewhat different from the BOF practices. As would be expected from the differences in the direction of the oxygen blow in these top and bottom-blowing processes, concentrations of iron and manganese oxides in BOF slags will be higher than those in Q-BOP slags. For the same reason the slag temperature in BOF is higher than the metal temperature, the reverse being the case for Q–BOP.

### 8.4.3 SLAG COMPOSITION AT TURNDOWN

Composition of slags in low alloy steelmaking by the BOF and Q–BOP processes varies within the following range, depending on the carbon content of steel at tap: 40 to 60% CaO, 4 to 8% MgO, 3 to 8% MnO, 5 to 35% FeO (total), 10 to 28% $SiO_2$, 1 to 3% $P_2O_5$, 1 to 2% $Al_2O_3$, 0 to 2% $CaF_2$, 0.1 to 0.2% S and minor amounts of other oxides. As pointed out earlier, the sum of the oxides CaO + MgO + FeO + $SiO_2$ is in the range 88 to 92 percent.

As noted from the plant data in Fig. 8.6, for the turndown temperatures of 1625 ± 25°C, the total iron oxide content of the finishing slag at tap decreases with increasing concentrations of $SiO_2$ and CaO. The dotted lines are for the quaternary system CaO–MgO–FeO–$SiO_2$ saturated with dicalcium (magnesium) silicate at 1600°C (Fig. 5.12). The readings from Fig. 5.12 for the silicate saturated melts are readjusted to CaO + MgO + FeO + $SiO_2$ = 90%; these are depicted by the dotted lines in Fig. 8.6. There is a pronounced shift in the position of the univariant equilibrium in the composition diagram for the saturated-steelmaking slags, to higher concentrations of CaO and to the corresponding lower concentrations of $SiO_2$, as compared to the quaternary system CaO–MgO–FeO–$SiO_2$ in equilibrium with liquid iron. This change in composition of the saturated slags to higher concentrations of CaO is due, in part, to slag temperatures being somewhat higher than 1600°C, also to the presence of $P_2O_5$, MnO,

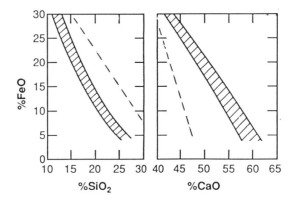

Fig. 8.6   The total iron oxide (as FeO) content of steelmaking slags is related to the SiO$_2$ and CaO contents. The dotted line is for CaO–6%MgO–FeO–SiO$_2$ system saturated with dicalcium (magnesium) silicate.

Fe$_2$O$_3$ and other minor oxides which increase the lime solubility in the slag.

The iron oxide content and slag basicity are related as shown in Fig. 8.7. The lower part of the hatched area is usually for BOF slags and the upper part for Q–BOP slags.

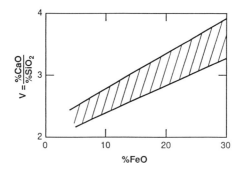

Fig. 8.7   Variation of slag basicity with total iron oxide content of steelmaking slags containing P$_2$O$_5$ < 3%.

## 8.5 DECARBURISATION

In most oxygen steelmaking processes the rate of oxygen blowing is usually within the range 2.2 to 3.0 Nm$^3$ min$^{-1}$ t$^{-1}$, depending on the composition of the furnace charge. During the first three to four minutes of oxygen blowing most of the silicon is oxidised, during which time about 0.4 or 0.5%C of the carbon is oxidised with an increasing rate of decarburisation as shown schematically in Fig. 8.8 for BOF heats. For Q–BOP heats the decarburisation rate curve has a similar shape, the only difference being in the duration of oxygen blowing which is about 16 to 18 minutes. Depending on the rate of oxygen blowing the plateau of the maximum rate of decarburisation is in the range 0.20 to 0.28%C per minute, which corresponds to the consumption of about 90 percent or more of the oxygen blown per minute.

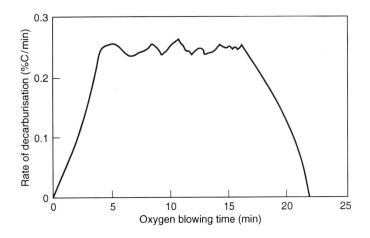

Fig. 8.8   Schematic representation of change in the decarburisation rate during oxygen blowing in BOF steelmaking..

Below about 0.4%C, the rate of decarburisation decreases with a decreasing carbon content because of an increase in the consumption of oxygen by the oxidation of phosphorus, manganese, iron and an increasing amount of oxygen dissolution in the steel bath. As an example let us assume that, starting at 17 minutes of the blowing time with the steel containing 0.4%C, the rate of decarburisation will decrease linearly with a decreasing carbon content of the steel bath during the last five minutes of the blow, for which the rate equation will be

$$\frac{d\%C}{dt} = -(\%C - \%C_{eq})\,(t - 17) \tag{8.4}$$

where $\%C_{eq}$ is the aimed equilibrium carbon content according to the charge-control model. Upon integration

$$\ln \frac{\%C - \%C_{eq}}{\%C_o - \%C_{eq}} = (t - 17) \tag{8.5}$$

Inserting $\%C_o = 0.4$ at t = 17 min and $\%C_{eq} = 0.04$

$$\ln \frac{\%C - 0.04}{0.036} = (t - 17) \tag{8.5a}$$

For the case considered as an exmaple, the decrease in the carbon content and increase in the oxygen content of the bath would be as given below.

| *t*, min | %C | ppm O* |
|---|---|---|
| 17 | 0.40 | 75 |
| 18 | 0.172 | 174 |
| 19 | 0.089 | 337 |
| 20 | 0.058 | 517 |
| 21 | 0.047 | 638 |
| 22 | 0.042 | 714 |

## 8.6 COMBINED-BLOWING IN BOF STEELMAKING

There are various types of combined-blowing practices in BOF steelmaking; the major ones are listed in Table 8.1.

**Table 8.1**   Major combined-blowing practices

| Process | Developed by | Bottom Gases injected | Flow rate $Nm^3 \, min^{-1} \, t^{-1}$ | Bottom wear rate mm per heat |
|---|---|---|---|---|
| LBE | ARBED-IRSID[6] | $N_2$, Ar | 0.01–0.10 | 0.3–0.6 |
| LD–CB | Nippon Steel Corp.[7] | $CO_2$, $N_2$, Ar | 0.02–0.06 | 0.3–0.8 |
| LD–KGC | Kawasaki Steel Corp.[8] | CO, $N_2$, Ar | 0.01–0.20 | 0.2–0.3 |
| LD–OTB | Kobe Steel Corp. Ref.[9,10] | CO, $N_2$, Ar | 0.01–0.10 | 0.2–0.3 |
| NK–CB | Nippon Kokan K.K.[11] | $CO_2$, $N_2$, Ar | 0.02–0.10 | 0.7–0.9 |

A brief description is given here only of the LBE system: incidentally, LBE is an acronym for 'Lance Bubbling Equilibrium'. In this system there are 10 to 16 gas injection elements arranged in one or two concentric circles on the converter bottom as shown in Fig. 8.9a. Nitrogen or argon is fed to the valve stand at a pressure of 10 atm and injected through the bottom of the vessel at gas flow rates of up to 0.1 $Nm^3min^{-1}t^{-1}$. The injection elements used may be of type I or II shown in Fig. 8.9b. In type I, the gas is blown through copper channels that are inserted in the grooved refractory plate encased in a stainless steel shell; in type II, 2 mm inside-diameter stainless steel tubes are embedded in a one-piece MgO–C refractory block.

*For the product [%C][ppm O] = 30, average in 200–240 t BOF heats.

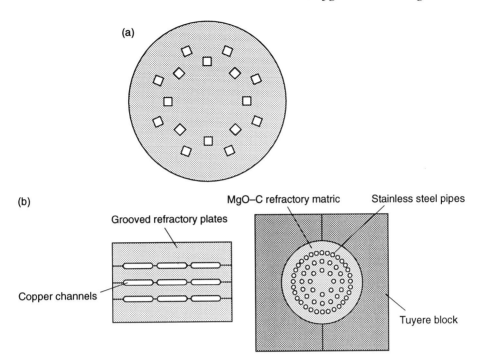

Fig. 8.9   LBE bottom gas injection devices.

Operation of the LBE bottom-stirring system requires that the furnace bottom be slagged to protect the refractories from excessive erosion. The thickness of the slag layer should be between 50 and 100 mm, maintained within the desired range by daily bottom measurements with the AGA laser.

There are essentially two gas bottom-stirring practices as outlined in Fig. 8.10 and 8.11 for low carbon (<0.1%) and medium or high carbon (>0.1%) heats. For low carbon heats the rate of nitrogen flow is low during the (i) waiting period, (ii) furnace charge and (iii) first ten minutes of the oxygen blow. Then argon is substituted for nitrogen at a low flow rate. During the last three minutes of the blow the argon flow rate is increased and changed back to the low flow rate at turndown for sampling. Post-bottom stirring is done with a high argon flow rate for less than two minutes. Finally a high nitrogen flow is used during the slagging operation. For high carbon heats the gas is maintained at a low flow rate throughout the blow.

As noted from the LD plant data for 315 t heats in Fig. 8.12 (cited by Deo and Boom[12]), the product [%C][ppm O] is 33 ± 2 for the normal LD practice. In the combined blowing practice this product is reduced to 21 ± 2 because of the dilution of CO in the gas bubbles. A similar behaviour is seen from the plant data in Fig. 8.13 for BOP and LBE–BOP practices. A

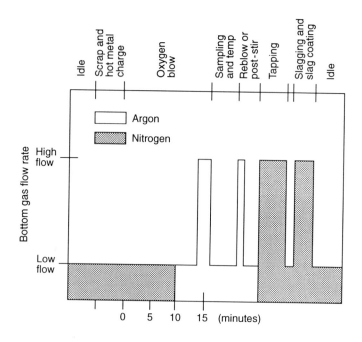

Fig. 8.10    Bottom-stirring practice for low-carbon heats (< 0.10%C).

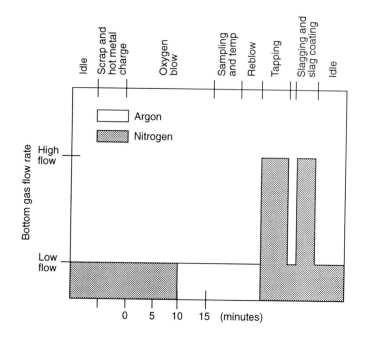

Fig. 8.11    Bottom-stirring practice for medium- and high-carbon heats (> 0.10%C).

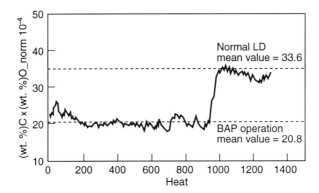

Fig. 8.12   Average values of the product [%C][%O] at turndown with normal LD operation and combined-blowing (BAP) operation. From Ref. 12.

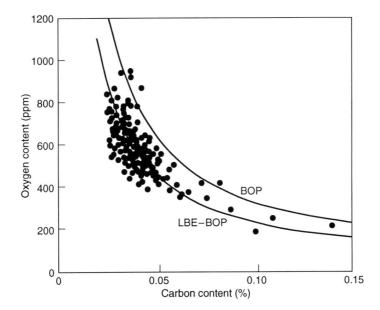

Fig. 8.13   Turndown carbon and oxygen contents in BOP steelmaking with and without bottom gas injection.

lower level of oxygen in the combined-blowing practice is also reflected on a slightly higher residual manganese in the steel at turndown and lower iron oxide content in the slag. Another advantage is that with combined blowing the heat can be blown to a lower level of carbon, e.g. 0.025 to 0.03%C, without over oxidising the steel bath.

There is also the possibility of removing some carbon and oxygen during post-bottom stirring with argon by the reaction

$$[C] + [O] \rightarrow CO(g) \tag{8.6}$$

If the kinetics of the chemical reaction at the melt-bubble surface is favour-able and the rate is controlled by the mass transfer of C and O to the bubble surface, the rate equation would be as given in section 2.6.4

$$\ln \frac{\%C - \%C_{eq}}{\%C_o - \%C_{eq}} = -S_o k_m t \tag{8.7}$$

where $t$ is the duration in seconds of post-bottom argon injection, $\%C_o = 0.04$ at turndown ($t = 0$) and $\%C_{eq}$ is for the C–O equilibrium for an average CO partial pressure in the gas bubbles. For the argon injection rate of 0.1 $Nm^3min^{-1}t^{-1}$, the volume flow rate of gas bubbles in the melt (for 1600°C and 1.5 atm average pressure) would be $\dot{V}_T = 0.457$ m³min$^{-1}$t$^{-1}$. For 220 t heat, $\dot{V}_T = 1.68$ m³s$^{-1}$, for which the previously derived rate constant from Fig. 2.32 is $S_o k_m = 0.015$ s$^{-1}$. Inserting these values in equation (8.7) gives

$$\ln \frac{\%C - \%C_{eq}}{0.04 - \%C_{eq}} = -0.015\, t \tag{8.8}$$

With the equilibrium constant [%C][ppm O] = 20 × $p_{CO}$ (atm), the values of $\%C_{eq}$ are derived for every incremental time of argon injection from C and O contents in the melt. Results of the calculations are given below.

| $t$, seconds | %C | $\%C_{eq}$ | ppm O | $p_{CO}$ atm* |
|---|---|---|---|---|
| 0 | 0.0400 | | 800 | |
| | | 0.026 | | 0.957 |
| 30 | 0.0353 | | 737 | |
| | | 0.025 | | 0.870 |
| 60 | 0.0316 | | 688 | |
| | | 0.024 | | 0.765 |
| 90 | 0.0288 | | 651 | |
| | | 0.023 | | 0.724 |
| 120 | 0.0263 | | 618 | |

These calculated rates of decarburisation are about twice those observed in trial heats at Gary BOP Shop of U.S. Steel, reported by Wardrop.[13] Evidently the rate of reaction in equation (8.6) is not controlled solely by mass transfer. Because of the retarding effects of oxygen and sulphur on the kinetics of the C–O reaction, the rate of decarburisation during the post-bottom argon injection will be controlled jointly by the interfacial chemical reaction and the mass transfer of C and O to the bubble surface.

## 8.7 METAL EMULSION IN FOAMING SLAGS

A foam is a heterogeneous medium consisting of gas bubbles dispersed in a liquid. Similarly, an emulsion is a dispersion of finely sub-divided

---

*For an incremental time $\Delta t$ with $n_{Ar}$ mol Ar injection, $n_{CO}$ mol CO generation and for an average bubble pressure of 1.5 atm, $p_{CO} = \{n_{CO}/(n_{CO} + n_{Ar})\}$ 1.5 atm.

particles of a liquid in an immiscible liquid. The formation of foams and emulsions and their relative stabilities are strongly influenced by surface tension and viscosity. The energy required to create foam or emulsion increases with an increasing surface tension of the liquid. The foam stability is determined by the rate of drainage of the liquid film between the bubbles; the lower the viscosity the faster the rate of drainage. Consequently, liquids of high surface tension and low viscosity, such as metals and mattes, cannot foam. Because molten slags have low surface tensions and high viscosities, they are susceptible to foaming in processes involving extensive rapid gas evolution as in oxygen steelmaking.

The foam is unstable; when the gas supply is terminated the foam collapses under the influence of interfacial tension (the Marangoni effect) and gravity force. The lower the surface tension and higher the viscosity, the slower would be the rate of collapse of the foam. Cooper and Kitchener[14] have studied the foaming characteristics of metallurgical slags. They found that the relative foam stability increases with a decreasing temperature and increasing $SiO_2$ and $P_2O_5$ contents of the slag. Swisher and McCabe[15] observed similar effects of the additions of $B_2O_3$ and $Cr_2O_3$ to silicate melts. When the surface tension/viscosity ratio $(\sigma/\eta)$ is lowered by changing temperature and/or slag composition, the relative foam stability increases.

A large volume of CO evolved at a high rate from the steel bath during decarburisation* causes ejection of steel droplets into the slag layer. The extent of metal emulsion in slag in pneumatic steelmaking processes was first brought to light by Kozakevitch and co-workers.[16–18] From 200 t BOF trial heats, Meyer et al.[19] and Trentini[20] found that about 30 percent of the metallic charge is dispersed in the slag during the period of a high rate of decarburisation. The metal droplets are in the range 0.1 to 5 mm diameter, and the slag-metal interfacial area has been estimated to be 100 to 200 $m^2$ per tonne of steel in the slag layer. According to the studies of Meyer et al., the average residence time of metal droplets in the slag is about 1 to 2 minutes, the rate of decarburisation in the metal emulsion is 0.3 to 0.6%C per minute and in the entire system about 0.25% C per minute. Near the end of decarburisation, the slag foam subsides and most of the metal droplets return to the steel bath.

## 8.8 CONTROL OF SLAG FOAMING

Excessive slag foaming during high rate of decarburisation causes bath slopping resulting in an overflow of slag from the vessel mouth. Several

---

*At the high rate of decarburisation 0.25% C $min^{-1}$, the volume of CO generated at 1600°C and 1.5 atm average bubble pressure would be 4699 $m^3$ $min^{-1}$ in a 220 tonne heat; the superficial gas velocity being about 3.6 m $s^{-1}$.

devices are being used to check the slag height in the vessel and to make necessary changes in the blowing practice to curtail bath slopping. A sound detecting system has been used as an indicator of the onset of slopping. The main part of this system is a condenser microphone which is contained in a water cooled probe and placed in a dust free location in the hood.

The control system described by Iida *et al.*[1] consists of (i) a device to measure lance vibration and (ii) a probe attached to the sublance to measure slag height. These measurements together with lance height and oxygen flow rate are put into the computer control system using the following empirical relation.

$$S_H = \frac{G - b}{a\,V_o} + L_H + B_H \qquad (8.9)$$

where 
$G$ = acceleration of lance vibration, cm s$^{-2}$
$V_o$ = oxygen flow rate, Nm$^3$min$^{-1}$
$S_H$ = foaming slag height, m
$L_H$ = lance height, m
$B_M$ = furnace – bottom height correction, m
$a, b$ = constants.

The principle of this slag formation control system is shown in Fig. 8.14. When the judgment result is 'good', the blowing pattern continues as previously programmed. When the result is 'not good' the blowing programme is adjusted for soft or hard blow by changing the oxygen blow rate and/or lance height. After the judgment result shows 'good' the initial programme is then followed. An example of the automatic blowing practice to curtail slopping is shown in Fig. 8.15.

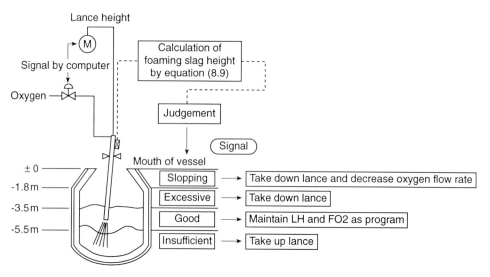

Fig. 8.14   Principle of automatic control of slag formation. From Ref. 1.

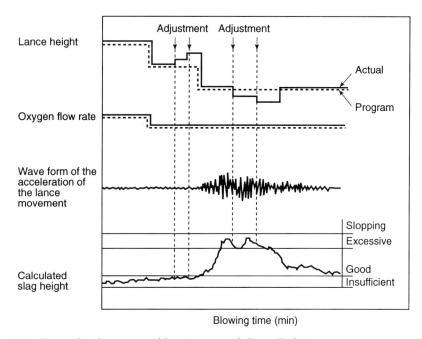

Fig. 8.15   Example of automatic blowing control. From Ref. 1.

Variation of the lance height pattern during the blow depends on the design of the oxygen-lance nozzle. The design features of *de Laval* nozzles used at some plants are summarised in Table 8.2, reproduced from the book by Deo and Boom.[12] Depending on the nozzle design and oxygen blowing rate, the blowing patterns are as shown schematically in Fig. 8.16 as given in Ref. 12.

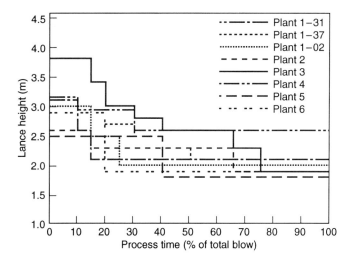

Fig. 8.16   Blowing regime of various plants; for details of plant names see footnote to Table 8.2. From Ref. 12.

Reference may be made also to a review paper by Koria[21] on the fundamentals of nozzle design for top-blown oxygen steelmaking and empirical correlations for blowing regimes in relation to the converter capacity.

**Table 8.2**   Lance head design data of some steel plants. From Ref. 12

|  |  | Plant 1 #31 | Plant 1 #37 | Plant 1 #02 | Plant 2 | Plant 3 | Plant 4 | Plant 5 | Plant 6 |
|---|---|---|---|---|---|---|---|---|---|
| $F_{O_2}$ | (m³min⁻¹) | 915 | 915 | 660 | 650 | 750 | 900 | 900 | 330 |
| $p_o$ | (Pa) | 10.2 | 10.8 | 6.6 | 11.1 | 13.3 | 11.6 | 11.0 | 7.9 |
| $d_t$ | (mm) | 46.6 | 45.2 | 55.0 | 37.6 | 36.9 | 48.44 | 40.6 | 41.0 |
| $d_e$ | (mm) | 58.0 | 61.3 | 65.75 | 55.35 | 51.95 | 63.0 | 57.75 | 50.5 |
| $\theta_1$ | (deg) | 12 | 14 | 10 | 14 | 16 | 12 | 14 | 10 |
| $n$ |  | 5 | 5 | 4 | 5 | 5 | 4 | 6 | 3 |
| $W_{bath}$ | (tons) | 315 | 315 | 315 | 245 | 230 | 330 | 300 | 100 |
| Bottom-stirring system |  | BAP | BAP | BAP | tuyères φ 5.5 mm | LBE | BAP | – | BAP |
| $Si_{hot\ metal}$ | (%) | 0.400 | 0.400 | 0.400 | 0.330 | 0.400 | 0.750 | 0.600 | 0.400 |

Notes: Plant information:
  Plant 1: Hoogovens IJmuiden, BOS No. 2,
        #31: lance head design 31, converter 23;
        #37: lance head design 37, converter 23;
        #02: lance head design 02, converter 21 and 22.
  Plant 2: BS Lackenby.
  Plant 3: Sollac Dunkerque.
  Plant 4: BS Port Talbot.
  Plant 5: BS Scunthorpe.
  Plant 6: Hoogovens IJmuiden, BOS No. 1.

## 8.9 STATES OF STEELMAKING REACTIONS AT THE END OF OXYGEN BLOWING*

Over the years various studies have been made of the plant analytical data on slag and metal samples taken from the vessel at turndown in an attempt to assess the states of slag–metal reactions at the end of oxygen blowing. Invariably there were variations in the results of these independent studies and the conclusions drawn from them with regard to the state of steelmaking reactions.

The direction of oxygen blowing, either by top lancing as in BOP (LD, BOP etc.) or bottom injection together with powdered lime through the bath as in Q–BOP, has a pronounced effect on the states of steelmaking reactions at the end of the blow. The analysis of both the BOF and Q–BOP plant data should give a better insight to the understanding of the state of gas–slag–metal reactions in oxygen steelmaking processes. The analytical

*This subject was discussed by the author in a recent publication.[22]

plant data used in this study were on samples taken from the vessel at first turndown from the BOP and Q–BOP shops of U.S. Steel. These plant data were acquired through the kind collaboration of the USS research personnel at the Technical Center.

8.9.1 OXYGEN–CARBON RELATION

Noting from the plot in Fig. 8.17, there is an interesting correlation between the product [ppm O] [%C] and the carbon content of the steel. The square root correlation holds up to about 0.05%C in BOP and up to 0.08% C in Q–BOP steelmaking. At low carbon contents, the oxygen content of steel in the BOP practice is higher than that in Q–BOP steelmaking. For carbon contents above 0.15% the product [ppm O] [%C] is essentially

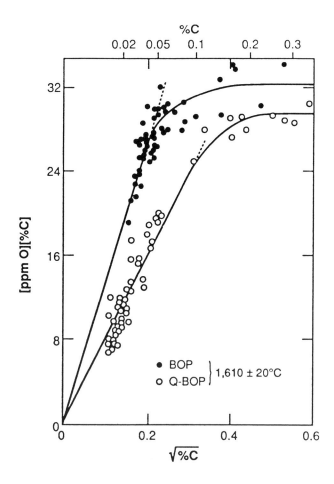

Fig. 8.17   Variation of product [ppm O][%C] with carbon content of steel at first turndown.

constant at about 30 ± 2, which is the equilibrium value for an average gas (CO) bubble pressure of about 1.5 atmosphere in the steel bath.

At low carbon levels in the melt near the end of the blow, much of the oxygen is consumed by the oxidation of iron, manganese and phosphorus, resulting in a lower volume of CO generation. With the bottom injection of argon in the BOP combined-blowing practice, and the presence of hydrogen in the gas bubbles in Q–BOP, the partial pressure of CO in the gas bubbles will be lowered in both processes when the rate of CO generation decreases. A decrease in the CO partial pressure at low carbon contents will be greater in Q–BOP than in BOP practices, because the hydrogen content of gas bubbles in Q–BOP is greater than the argon content of gas bubbles in BOP. It is presumably for this reason that the concentration product [O][C] in Q–BOP steelmaking is lower than in the BOP combined-blowing practice, particularly at low carbon levels.

The non-equilibrium states of the carbon–oxygen reaction at low carbon contents in BOP and Q–BOP are represented by the following empirical relations.

$$\text{BOP with } \underline{C < 0.05\%}:$$
$$[\text{ppm O}] \sqrt{\%C} = 135 \pm 5 \tag{8.10}$$

$$\text{Q-BOP with } \underline{C < 0.08\%}:$$
$$[\text{ppm O}] \sqrt{\%C} = 80 \pm 5 \tag{8.11}$$

### 8.9.2 IRON OXIDE–CARBON RELATION

There is a striking difference in the iron oxide–carbon relation between BOP and Q–BOP steelmaking. At all levels of turndown carbon, the iron oxide content of the BOP slag is about twice that of the Q–BOP slag (Fig. 8.18). The dotted line depicts the slag–metal equilibrium value of $(\%FeO)$ $[\%C] \approx 1.25$ given in equation (6.6).

As depicted in Fig. 8.19, the square root correlation also applies to the product $(\%FeO)[\%C]$ for low carbon contents, with marked departure from this empirical correlation at higher carbon contents. The slopes of the lines for low carbon contents are given below.

$$\text{BOP with } \underline{C < 0.10\%}:$$
$$(\%FeO) \sqrt{\%C} = 4.2 \pm 0.3 \tag{8.12}$$

$$\text{Q-BOP with } \underline{C < 0.10\%}:$$
$$(\%FeO) \sqrt{\%C} = 2.6 \pm 0.3 \tag{8.13}$$

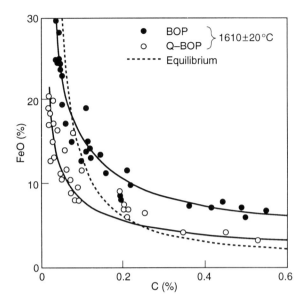

Fig. 8.18   Iron oxide–carbon relations in BOP and Q–BOP are compared with the average equilibrium relation (– – – –).

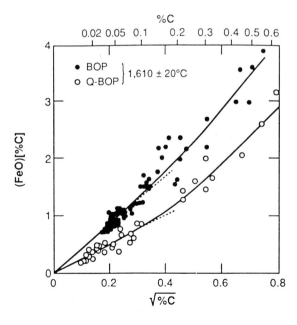

Fig. 8.19   Variation of product (FeO)[%C] with carbon content of steel at first turndown.

### 8.9.3 MANGANESE OXIDE-CARBON RELATION

At low carbon contents the ratio [%Mn]/(%MnO) may also be considered to be essentially proportional to $\sqrt{\%C}$ as shown in Fig. 8.20. The following are the slopes of the lines for the empirical non-equilibrium relations.

BOP with C < 0.10%:

$$\frac{[\%Mn]}{(\%MnO)} \frac{1}{\sqrt{\%C}} = 0.1 \pm 0.02 \tag{8.14}$$

Q-BOP with C < 0.10%:

$$\frac{[\%Mn]}{(\%MnO)} \frac{1}{\sqrt{\%C}} = 0.2 \pm 0.02 \tag{8.15}$$

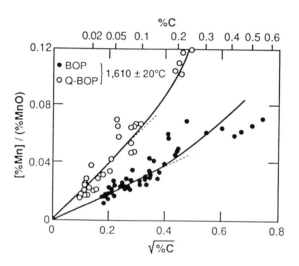

Fig. 8.20   Metal/slag manganese distribution ratios in BOP and Q–BOP are related to carbon content of steel at first turndown.

### 8.9.4 FeO–MnO–Mn–O RELATIONS

From the foregoing empirical correlations for the non-equilibrium states of reactions involving the carbon content of steel, the relations obtained for the reaction of oxygen with iron and manganese are compared in the table below with the equilibrium values for temperatures of 1610 ± 20°C and slag basicities of B = 3.2 ± 0.6.

|  | BOP $C < 0.05\%$ | Q-BOP $C < 0.08\%$ | Values for slag-metal equilibrium |
|---|---|---|---|
| $\dfrac{[\text{ppm O}]}{(\%\text{FeO})}$ | $32 \pm 4$ | $32 \pm 5$ | $26 \pm 9$ |
| $\dfrac{[\%\text{Mn}][\text{ppm O}]}{(\%\text{MnO})}$ | $13.6 \pm 3.2$ | $16.1 \pm 2.6$ | $18 \pm 6$ |
| $\dfrac{(\text{MnO})}{(\%\text{FeO})[\%\text{Mn}]}$ | $2.6 \pm 1.0$ | $2.1 \pm 0.6$ | $2.2 \pm 0.4$ |

It is seen that the concentration ratios of the reactants in low carbon steel describing the states of oxidation of iron and manganese, are scattered about the values for the slag–metal equilibrium. However, as indicated by the plant data in Fig. 8.21, the oxidation of iron and manganese are in the non-equilibrium states for high carbon contents in the steel at turndown. Although the concentration product [O][C] is close to the equilibrium value for an average CO pressure of about 1.5 atm in the steel bath, the concentrations of iron oxide and manganese oxide in the slag are above the equilibrium values for high carbon contents in the melt.

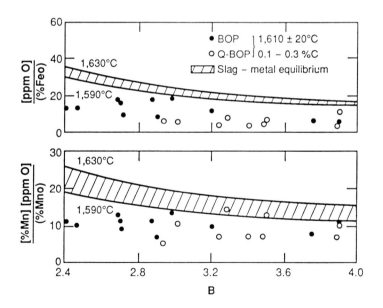

Fig. 8.21   Non-equilibrium states of oxidation or iron and manganese at high carbon contents in steel.

It is concluded from these observations that
(i)   at low carbon contents the equilibrium state of iron and manganese oxidation controls the concentration of dissolved oxygen

(ii) at high carbon contents it is the CO–C–O equilibrium which controls the oxygen content of the steel.

### 8.9.5 STATE OF PHOSPHORUS REACTION

The slag/metal phosphorus distribution ratios at first turndown in BOP and Q–BOP steelmaking are plotted in Fig. 8.22 against the carbon content of the steel. The plant data are for the turndown temperatures of 1610 ± 20°C and slags containing 50 ± 2% CaO and 6 ± 2% MgO. The shaded area is for the slag-metal equilibrium for the above stated conditions in Q–BOP, based on the empirical [O][C] relations, i.e. for C < 0.08%, [ppm O] $\sqrt{\%C}$ = 80 and for C > 0.15%, [ppm O][%C] = 30.

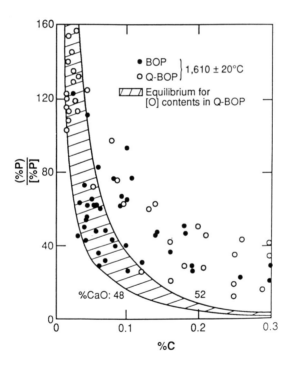

Fig. 8.22   Slag/metal phosphorus distribution ratios at first turndown in BOP and Q–BOP practices; slag–metal equilibrium values are within the hatched area for Q–BOP.

Noting that the oxygen contents of the steel at low carbon levels are greater in BOP steelmaking, the equilibrium phosphorus distribution ratios will likewise be greater in BOP than in Q–BOP steelmaking. For example, at 0.05% C and about 600 ppm O in BOP at turndown the average equilibrium value of (%P)/[%P] is about 200 at 1610°C, as compared to the average value of 60 in Q–BOP at 0.05% C with about 360 ppm O.

Below 0.04% C the phosphorus distribution ratios in Q–BOP are in general accord with the values for slag–metal equilibrium. However, at higher carbon contents the ratios (%P)/[%P] are well above the equilibrium values. In the case of BOP steelmaking below 0.1% C at turn-down, the ratios (%P)/[%P] are much lower than the equilibrium values. On the other hand, at higher carbon contents the phosphorus distribution ratios are higher than the equilibrium values as in the case of Q–BOP.

The effect of temperature on the phosphorus distribution ratio in Q-BOP is shown in Fig. 8.23 for melts containing 0.014% to 0.022% C with BO = 52 ± 2% in the slag.

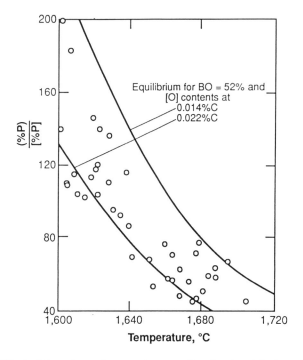

Fig. 8.23   Effect of turndown temperature on the slag/metal phosphorus distribution ratios in Q–BOP for turndown carbon contents of 0.014 to 0.022% C; curves are for slag–metal equilibrium.

8.9.6 STATE OF SULPHUR REACTION

A highly reducing condition that is required for extensive desulphurisation of steel is opposite to the oxidising condition necessary for steel making. However, some desulphurisation is achieved during oxygen blowing for decarburisation and dephosphorisation. As seen from typical examples of the BOP and Q–BOP plant data in Fig. 8.24, the state of steel desulphurisation at turndown, described by the expression [%O](%S)/

[%S], is related to the SiO$_2$ and P$_2$O$_5$ contents of the slag. Most of the points for Q–BOP are within the hatched area for the slag-metal equilibrium reproduced from Fig. 6.9. However, in the case of BOP (or BOF) steelmaking the slag/metal sulphur distribution ratios at turndown are about one-third or one-half of the slag-metal equilibrium values. At higher carbon contents, e.g. C>0.1%, the sulphur distribution ratios in both processes are below the slag-metal equilibrium values.

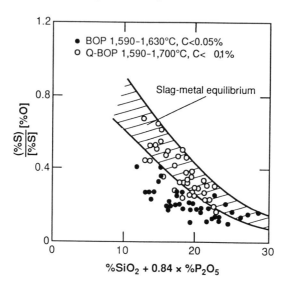

Fig. 8.24   Equilibrium and non-equilibrium states of sulphur reaction in Q–BOP and BOP (or BOF) steelmaking.

### 8.9.7 HYDROGEN AND NITROGEN CONTENTS IN BOF AND Q–BOP STEELMAKING

Hydrogen and nitrogen are removed from the steel bath during oxygen blowing by CO carrying off H$_2$ and N$_2$.

$$2\underline{H} \text{ (in steel)} \rightarrow H_2 \text{ (g)} \qquad (8.16)$$

$$2\underline{N} \text{ (in steel)} \rightarrow N_2 \text{ (g)} \qquad (8.17)$$

There is however continuous entry of both hydrogen and nitrogen into the bath by various means. In both BOF and Q–BOP there is invariably some leakage of water from the water cooling system of the hood into the vessel. In the case of Q–BOP, the natural gas (CH$_4$) used as a tuyere coolant is a major source of hydrogen.

The hydrogen content of steel in the tap ladle, measured by a probe called HYDRIS, is less than 5 ppm H in BOF steelmaking and 6 to 10 ppm H in Q–BOP steelmaking. Because natural gas (CH$_4$) is used as a tuyere coolant in the Q–BOP the hydrogen content of steel made in this vessel is always higher than that in the BOF vessel. In both practices the re-blow will always increase the hydrogen content of the steel. A relatively small volume of CO evolved during the re-blow cannot overcome the hydrogen pickup from various sources.

In the BOF practice the steel in the vessel contains 12 to 18 ppm N at first turndown. In the Q–BOP practice the nitrogen content of the steel in the vessel is in the range 15 to 25 ppm at first turndown. When there is a re-blow the nitrogen contents of the steel in both processes will be higher than those at the first turndown.

The primary source of nitrogen in Q–BOP steelmaking is the natural gas which contains about 1% $N_2$. It has long been known that in BOF steel-making the quantity of oxygen consumed in all the reactions during the blow is always a little higher than the quantity that is lanced into the melt. This difference is a positive indication of air entrainment into the oxygen jet stream in toplancing, hence some nitrogen pickup by the steel.

### 8.9.8 GENERAL COMMENTS

Since the plant data cited were acquired from the BOP shops of U.S. Steel, discussion on the oxygen top blowing process has been abbreviated as BOP steelmaking. Observations made on the states of reactions at turndown from the BOP plant data apply equally well to all other types of top blowing practices, known under different names, e.g. LD, BOS, BOF . . . etc.

The analyses of plant data on slag and metal samples taken at first turndown have revealed that there are indeed equilibrium and non-equilibrium states of reactions in BOF and Q-BOP steelmaking. In all the reactions considered, the carbon content of steel is found to have a deci-sive effect on the state of slag-metal reactions.

Considering the highly dynamic nature of steelmaking with oxygen blowing and the completion of the process in less than 25 minutes in BOF and less than 18 minutes in Q–BOP steelmaking, it is not surprising that the slag–metal reactions are in the non-equilibrium states in heats with high carbon contents at turndown. With regard to low carbon heats all the slag–metal reactions are close to the equilibrium states in Q–BOP. In the case of BOF steelmaking however, the states of steel dephosphorisation and desulphurisation are below the expected levels for slag-metal equilibrium. As we all recognise it is of course the bottom injection of lime, together with oxygen, that brings about a closer approach to the slag-metal equilibrium in Q–BOP as compared to the BOF practice, particularly in low carbon heats.

In the computer-based packages related to steelmaking processes, some of which were cited in sections 1.13 and 2.8 of Chapters 1 and 2 respectively, the formulations are invariably based on the assumption that the gas–slag–metal equilibrium prevails during oxygen blowing. It is all too clear from the assessment of the plant data presented here that due consideration must be given to the non-equilibrium states of steelmaking reactions, particularly in the BOF practice, in the formulation of the computer software on steelmaking processes.

## 8.10 STATES OF SLAG-METAL REACTIONS IN EAF STEELMAKING

Practical aspects of steelmaking and the plant equipment used in the electric arc furnace (EAF) have been adequately described and documented in detail in a series of papers compiled in a reference book entitled *Electric furnace steelmaking*, edited by Taylor[23]. Therefore this subject will not be discussed further. However, the subject barely touched upon in any EAF related papers is on the assessment of the states of slag–metal reactions at the time of furnace tapping. This lack of technical information on EAF steelmaking reactions is due mainly to a traditional reluctance of the EAF steelmaking plants to release from their heat logs the analytical data on slag and steel compositions which are encountered in the production of a wide variety of steel grades. It is hard to comprehend this attitude of the EAF steelmakers; it appears as though they have highly secretive recipes. After all, in the present practice of EAF steelmaking supplemented with oxygen lancing throughout the scrap melting and steel refining operations, the states of slag–metal reactions will hardly differ from what occurs in BOF or Q–BOP steelmaking, for which there are many publications based on plant data.

An estimate made in this section of Chapter 8 on the states of slag–metal reactions at tap is based on the analytical data of about fifty heats, which were given to the author from one of the EAF steelmaking plants for an assessment of the states of steelmaking reactions. The data considered here are for the grades of steel containing 0.05 to 0.20% C, 0.1 to 0.3% Mn, 0.1 to 0.3% Cr and small contents of Cu, Ni and Mo; the tap temperatures being in the range 1610 to 1640°C.

### 8.10.1 SLAG COMPOSITION

The iron oxide, silica and calcium oxide contents of slags are within the hatched areas shown in Fig. 8.25. The silica contents are slightly higher than those in the BOF or Q–BOP slags. On the other hand the CaO contents of EAF slags are about 10% lower than those in the BOF slags. This difference is due to the higher concentrations of $Al_2O_3$, $Cr_2O_3$ and $TiO_2$ in the EAF slags; 4 to 12% $Al_2O_3$, 1 to 4% $Cr_2O_3$ and 0.2 to 1.0% $TiO_2$. In these slags the basicity ratio B is about 2.5 at 10% FeO, increasing to about 4 at 40% FeO.

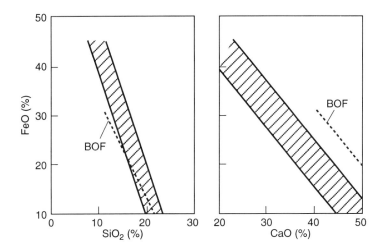

Fig. 8.25   FeO, SiO$_2$ and CaO contents of EAF slags at tap, are compared with the BOF slags.

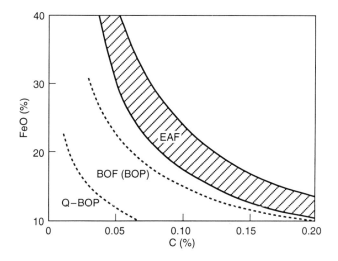

Fig. 8.26   Relation between %FeO in slag and %C in steel at tap in EAF steelmaking is compared with the relations in BOF and Q–BOP steelmaking.

8.10.2 IRON OXIDE–CARBON RELATION

As is seen from the plant data within the hatched area in Fig. 8.26, the iron oxide contents of EAF slags are much higher than those of the BOF slags for the same carbon content at tap. However, the product [%C][ppm O] at tap is about 26 ± 2 which is slightly lower than that for BOF or Q–BOP steelmaking.

8.10.3 FeO–MnO–Mn RELATION

For the slag-metal reaction involving [Mn], (FeO) and (MnO), the equilibrium relation $K'_{FeMn} = (\%MnO)/(FeO)[\%Mn]$ is 1.8 ± 0.2 at 1625 ± 15°C and the basicity B > 2.5, as shown in Fig. 6.2. The EAF data are scattered about $K'_{FeMn} = 1.8 ± 0.4$, which is near enough to the slag-metal equilibrium similar to those in BOF and Q–BOP steelmaking. The same conclusion was drawn from the results of trial heats in another EAF steelmaking plant, as shown later in Fig. 9.17.

8.10.4 STATE OF CHROMIUM OXIDATION

As shown earlier in Fig. 6.3, the slag/metal distribution ratio of chromium in both the electric arc furnace and open hearth furnace steelmaking, the latter now being obsolete, are in general accord with the slag–metal equilibrium values. That is, at tap the ratio $(\%Cr)/[\%Cr]$ is proportional to the iron oxide content of the slag, with a proportionality factor of 0.3 ± 0.1.

8.10.5 STATE OF PHOSPHORUS REACTION

For steels containing 0.07 to 0.09% C in the furnace at tap, the estimated dissolved oxygen contents will be about 370 to 290 ppm O. For slags containing (40 ± 2) %CaO and (6 ± 1) %MgO, the equilibrium constant $k_{PO}$ for the phosphorus reaction, given in equations (6.16) and (6.19), is in the range $2.57 \times 10^4$ to $4.94 \times 10^4$ at 1625°C. For the tap carbon contents and slag composition given above, the equilibrium slag/metal phosphorus distribution ratios will be in the range 4 to 13. The EAF plant data show ratios $(\%P)/[\%P]$ from 15 to 30. Similar to the behaviour in BOF steelmaking, the state of phosphorus oxidation to the slag in EAF steelmaking is greater than would be anticipated from the equilibrium considerations for carbon, hence oxygen contents of the steel at tap.

8.10.6 STATE OF SULPHUR REACTION

The state of steel desulphurisation, represented by the product $\{(\%S)/[\%S]\}$ $(\%FeO)$, decreases with increasing contents of $SiO_2$ and $P_2O_5$ in the slag, as shown in Fig. 8.27. The EAF plant data within the hatched area are below the values for the slag-metal equilibrium. This non-equilibrium state of the sulphur reaction in EAF steelmaking is similar to that observed in BOF steelmaking.

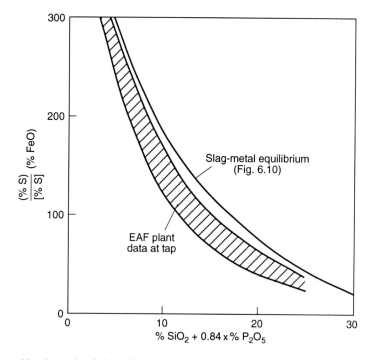

Fig. 8.27   Slag/metal sulphur distribution ratios at tap in EAF steelmaking is compared with the slag–metal equilibrium values.

8.10.7 SUMMARY

General indications are that the states of slag–metal reactions at tap in EAF steelmaking are similar to those noted in BOF and Q–BOP steelmaking at carbon contents above 0.05% C. That is, the slag/metal distribution ratio of manganese and chromium are scattered about the equilibrium values; the phosphorus and sulphur reactions being in non-equilibrium states. Also, the iron oxide content of the slag is well above the slag-metal equilibrium values for carbon contents of steel at tap. On the other hand, the product [%C][ppm O] = 26 ± 2 approximately corresponds to the C–O equilibrium value for gas bubble pressures of 1.3 ± 0.2 atm in the EAF steel bath.

REFERENCES

1. Y. IIDA, K. EMOTO, M. OGAWA, Y. MASUDA, M. ONISHI and H. YAMADA, *Kawasaki Steel Giho*, 1983, **15**, 126 (in Japanese); *Trans. Iron and Steel Inst. Japan*, 1984, **24**, 540.
2. A.I. VAN HOORN, J.T. VAN KONYNENBURG and P.J. KREYGER, in *Role of slag in basic oxygen steelmaking*, ed. W.–K. LU, p. 2–1, 1976, Hamilton, Ontario. McMaster University Press.

3. F. Bardenheurer, H. vom Ende and K.G. Speith, *Arch. Eisenhütenwes*, 1968, **39**, 571.

4. P. Nilles, E. Denis, F. Merken and P. Dauby, *CRM Metall, Rep.*, 1971, **27**, 3.

5. R. Baker, British Steel Corporation Report, code CAPL/SM/A/31/74.

6. F. Schleimer, R. Henrion, F. Goedert, G. Denier and J.C. Grosjean, *Iron and Steel Engineer*, 1981, **58**(12), 34.

7. H. Iso, Y. Jyono, K. Arima, M. Kanemoto, M. Okajima and H. Narita, *Trans. Iron and Steel Inst. Japan*, 1988, **28**, 49.

8. K. Emoto, T. Imai, F. Sudo, H. Take and T. Isoda, *Steelmaking Conf. Proc.*, 1987, **70**, 347.

9. S. Ito, M. Kitamura, S. Koyama, H. Matsui and H. Fujimoto, *Steelmaking Conf. Proc*, 1982, **65**, 123.

10. H. Yamana, T. Soejima, J. Kobayashi, H. Matsui, H. Fujimoto and N. Genma, *Steelmaking Conf. Proc.*, 1987, **70**, 339.

11. Y. Miyawaki, Y. Nimura, T. Usui, K. Yamada and T. Hirose, *Proc. 4th Process Technology Conf.*, 1984, 163.

12. B. Deo and R. Boom, *Fundamentals of steelmaking metallurgy*, Prentice Hall International, London, 1993.

13. R.M. Wardrop, *Steelmaking Conf. Proc.*, 1984, **67**, 107.

14. C.F. Cooper and J.A. Kitchener, *J. Iron and Steel Inst.*, 1959, **193**, 48.

15. J.H. Swisher and C.L. McCabe, *Trans. Met. Soc. AIME*, 1964, **230**, 1669.

16. P. Kozakevitch and P. Leroy, *Rev. Mét.*, 1954, **51**, 203.

17. P. Kozakevitch, G. Urbain, B. Denizot and H. Margot-Marette, *Congr. international sur les aciers a l'oxygene*, 1963, p. 248. Le Touget Inst. Res. Siderurgie.

18. P. Kozakevitch, *J. Metals*, 1969, **21**, 64.

19. H.W. Meyer, W.F. Porter, G.C. Smith and J. Szekely, *J. Metal*, 1968, **20**, 35.

20. B. Trentini, *Trans. Met. Soc. AIME*, 1968, **242**, 2377.

21. S.C. Koria, *Steel Res.*, 1988, **59**, 257.

22. *Ethem T. Turkdogan Symposium*, pp. 253–269, The Iron and Steel Society of AIME, May 15–17, 1994.

23. *Electric furnace steelmaking*, ed. C.R. Taylor. The Iron and Steel Society, Inc., 1985.

# CHAPTER 9

# Steel Refining in the Ladle

The treatment of steel in the ladle has gone through many facets of development during the past four decades. The first new major developments in the 1950s were ladle-to-ladle and ladle-to-mould vacuum degassing processes for hydrogen removal to prevent flake cracking in forgings and heavy-guage plate steels. In the 1960s more efficient vacuum degassing processes were developed: Dortmund–Hoerder (DH) and Ruhrstahl–Heraeus (RH), later modified to RH–OB by Nippon Steel Corporation. The advent of granulated flux injection into liquid steel and argon stirring for desulphurisation started in the 1970s, followed by the development of cored-wire feeding of elements for microalloying steel and inclusion morphology control. These innovations widened the scope of steel refining in the ladle to meet stringent requirements for high performance in continuous casting. The subjects to be discussed in this chapter constitute the major technical aspects of ladle refining processes which involve thermochemistry, gas–slag–steel reaction equilibria, reaction kinetics and fluid dynamics.

## 9.1 TAP LADLE

### 9.1.1 LADLE REFRACTORY LINING

In most steelmaking shops the ladle lining is made of high alumina bricks (70 to 80% $Al_2O_3$), with magnesia bricks for the slag line. The choice of magnesia bricks depends on the steel treatment in the ladle furnace (LF) and composition of the ladle slag. Magnesia bricks contain about 10% C and some metallic additives such as Mg, Al or Si to minimise the oxidation of carbon. References may be made to a paper by Schrader and Rankovic[1] for tests made of the slag attack on various commercial MgO–C bricks. With calcium aluminate slag containing 10% FeO and 3% $SiO_2$, the extent of erosion increases with relative brick flux content as shown in Fig. 9.1.

For a given ladle refining practice, the lining erosion, hence the ladle life, depends on the frequency of the ladle recycle and contact time. With infrequent ladle recycle, the thermal stresses on the lining increase causing a higher erosion rate. The erosion rate increases also with an increasing time of slag contact. The following are average numbers for ladle life, depending on the frequency of the recycle.

| Recycle, heats/ladle/day | Life, heats/ladle |
|---|---|
| 1 to 2 | 25 to 30 |
| 4 to 6 | 50 to 55 |

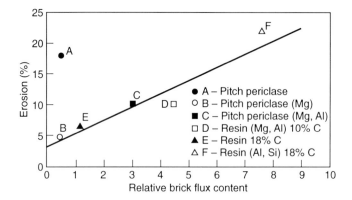

Fig. 9.1   Erosion by a 3-percent-$SiO_2$ slag vs. relative flux content. From Ref. 1.

In many electric furnace shops, the resin-bonded (calcined) dolomite brick lining is used for the ladles. Because the calcined dolomite refractories are sensitive to thermal cycling, cooling of the ladles is undesirable. Provided the ladle lining is not allowed to cool, the life of dolomite lined ladles is about twice that of the alumina brick lined ladles. Once the dolomite lined ladle is in service on a back-to-back cycle, preheating for only one hour is sufficient.

Stabilisation of calcined dolomite:

Upon calcination, the mineral dolomite, $CaCO_3 \cdot MgCO_3$, becomes a physical mixture of CaO and MgO.

$$CaCO_3 \cdot MgCO_3 \rightarrow CaO + MgO + 2CO_2 \qquad (9.1)$$

Because of the hydration of lime in air, the calcined dolomite is unstable, known as 'dusting' or 'perishing' in storage; the lower the calcination temperature, the greater the instability.

In order to stabilise the calcined dolomite, the lime must be converted to tri-calcium silicate during calcination. This is achieved by calcining a mixture of dolomite and the mineral serpentine so that the following reaction will occur during calcination.

$$\begin{array}{cc} \text{dolomite} & \text{serpentine} \\ 6(CaCO_3 \cdot MgCO_3) + 3MgO \cdot 2SiO_2 \cdot 2H_2O & \rightarrow \end{array}$$

$$2(3CaO \cdot SiO_2) + 9MgO + 12CO_2 + 2H_2O \qquad (9.2)$$

Typical composition of stabilised dolomite brick:
40.0% CaO, 40.3% MgO, 14.4% $SiO_2$, 3.5% $Fe_2O_3$ and 1.5% $Al_2O_3$.


9.1.2 LADLE PREHEATING

The basic refractory and high alumina bricks have high thermal conductivities (Fig. 9.2), therefore ladles must be preheated to avoid excessive

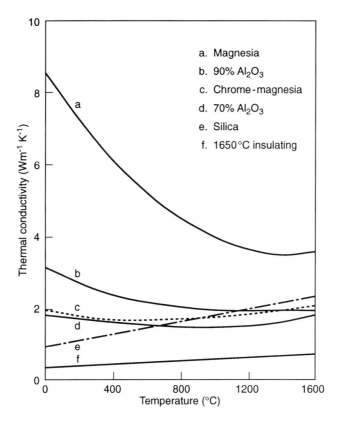

Fig. 9.2 Thermal conductivities of refractory bricks for ladle lining.

heat losses during furnace tapping and subsequent time of steel treatment in the ladle.

Many plant studies have been made of steel temperature losses in the ladle at various stages of processing and various aspects of ladle preheating. In a study made by Tomazin et al.[2] at LTV Steel Comp., temperature measurements were made of ladle lining at various positions for several preheating cycles. In addition, a full teemed heat cycle was monitored to establish temperatures achieved due to steel contact with the ladle lining. These data were used to fine-tune a computer model.

The hot surface of the ladle lining and shell temperature as a function of preheating time is shown in Fig. 9.3 for an initially cold dry ladle. Considerably longer time is required to preheat a newly relined ladle due to the moisture in the mortar and brick. The rate of rise of hot face temperature depends on the gap from the ladle top to the burner wall and the preheat thermal input. A rapid rate of heating should be avoided for the following reasons.

i.   Rapid heat up results in a non-equilibrium temperature profile, i.e. a steep temperature gradient below the hot face.

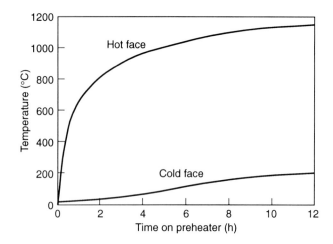

Fig. 9.3   Lining temperatures during preheating a cold dry ladle. From Ref. 2.

ii.  Rapid heat up generates extreme shell stresses.
iii. Thermal shock resistance of the brick may not be high enough to withstand a fast rate of heat up.

The change in temperature profile of a steel ladle lining is shown in Fig. 9.4 for the duration of preheating, transferring to the melt shop for tap and steel treatment. The hot face temperature decreases during transfer of the ladle from preheater to the melt shop. As shown in Fig. 9.5 computed by

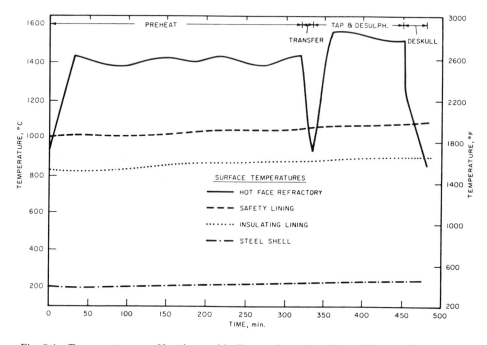

Fig. 9.4   Temperature profile of a steel ladle as a function of time. From Ref. 3.

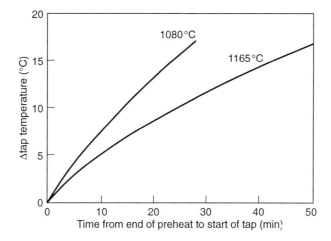

Fig. 9.5  The effect of cooling time for two different preheat temperatures for an uncovered ladle between end of preheat and tap, on the tap temperature adjustment. From Ref. 2.

Tomazin *et al.*, the tap temperature of the steel should be adjusted in accordance with the preheat hot face temperature and the time interval from end of preheat to start of tap.

Decrease in liquid steel temperature with time in the ladle is shown in Fig. 9.6 for ladle with or without a cover and with a slag layer. At the end of steel treatment, granular burnt rice hull is thrown onto the slag surface for additional thermal insulation to minimise heat losses from the steel during casting.

### 9.1.3 POROUS PLUG ON THE LADLE BOTTOM

The top lancing method is used for steel refining in the ladle with powder injection and stirring the melt with high argon flow rates. For gas bubbling at moderate rates, e.g. <0.6 $Nm^3min^{-1}$, a porous refractory plug is used. This is fitted usually on the bottom of the ladle as illustrated schematically in Fig. 9.7.

There are various designs of porous plugs; those in Fig. 9.8 are typical examples.[4] Reference may be made to a review paper by Anagbo and Brimacombe[5] on porous plugs in liquid metal refining in the ladle.

## 9.2 STEEL HOMOGENISATION WITH GAS STIRRING

The mixing of liquid steel with gas injection to achieve homogenisation of melt temperature and composition is due primarily to the dissipation of

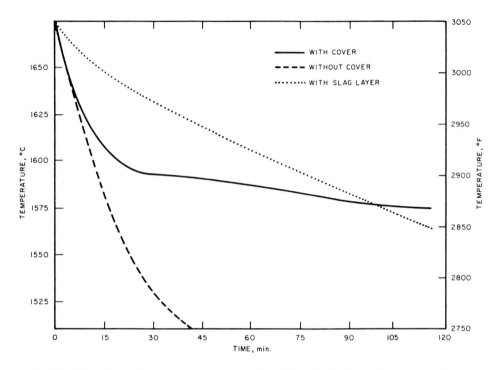

Fig. 9.6   Liquid-steel temperature as a function of time for ladles with cover, without cover, and with slag layer. From Ref. 3.

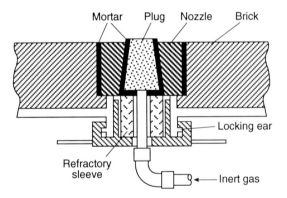

Fig. 9.7   Porous plug assembly in the bottom wall of a ladle. Plug life: 300 to 400 min stir time.

the buoyancy energy of the injected gas. The effective stirring power of gas is given by the following thermodynamic relation, which was derived by Pluschkell.[6]

$$\dot{\varepsilon} = \frac{\dot{n}RT}{M} \ln \frac{P_t}{P_o} \qquad (9.3)$$

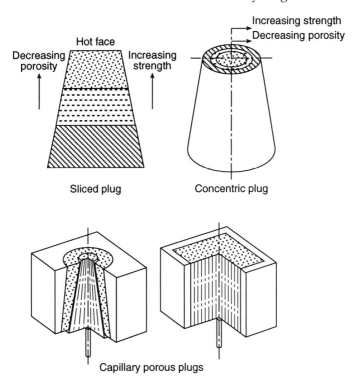

Fig. 9.8   Typical examples of porous plugs. From Ref. 4.

$\dot{\varepsilon}$ = stirring power, W t$^{-1}$
$\dot{n}$ = molar gas flow rate
$R$ = gas constant, 8.314 J mol$^{-1}$K$^{-1}$
$T$ = temperature, K
$M$ = mass of steel, t (tonne)
$P_o$ = gas pressure at the melt surface, atm
$P_t$ = total pressure at injection depth $H$ (m)
   = $P_o + \rho g H$
$\rho$ = steel density, 6940 kg m$^{-3}$ at ~ 1600°C
$g$ = 9.81 m s$^{-2}$

With these values the following equation is obtained.

$$\dot{\varepsilon}\ (Wt^{-1}) = 14.23\ \frac{\dot{V}T}{M}\ \log\left(1 + \frac{H}{1.48P_o}\right) \qquad (9.4)$$

where $\dot{V}$ = gas flow rate, Nm$^3$ min$^{-1}$.

### 9.2.1 MIXING TIME

The gas stirring time to achieve 95% homogenisation is called the mixing time $\tau$. Several experimental and theoretical studies have been made to formulate the mixing time in terms of the gas stirring power, ladle diameter and depth of gas injection in the melt. The following relation is from the work of Mazumdar and Guthrie.[7]

$$\tau(s) = 116(\dot{\varepsilon}\,)^{-1/3}(D^{5/3}H^{-1})\qquad(9.5)$$

where $D$ = average ladle diameter
$\quad\quad\;\; H$ = depth of gas injection

Calculated mixing times are given in Fig. 9.9 for the simplified case of $D \approx H$. In a 200 t heat, the melt homogenisation will be achieved with argon bubbling for 2.0 to 2.5 minutes at the rate of 0.2 Nm³ min⁻¹.

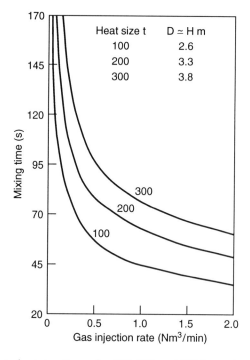

Fig. 9.9   Calculated mixing times for 100, 200 and 300 tonne heat sizes.

9.2.2 RATE OF SLAG–METAL REACTION IN GAS STIRRED MELTS

The rates of most slag–metal reactions are controlled primarily by mass transfer of the reactants and products across the slag–metal interface. In steel desulphurisation in the ladle, the slag–metal mixing is achieved with argon stirring, and the mass transfer controlled rate of desulphurisation is formulated as

$$\ln \frac{\%S - \%S_{eq}}{\%S_o - \%S_{eq}} = -kt\qquad(9.6)$$

where $S_o$ = initial content in steel,
$S_{eq}$ = content in equilibrium with the ladle slag,
$S$ = content at the end of argon stirring time,
$k$ = the rate constant, min$^{-1}$

From the pilot plant tests[8] with 2.5 t heats on steel desulphurisation, it was found that at moderate gas bubbling rates, corresponding to the power of stirring $\dot{\varepsilon}$ < 60 W t$^{-1}$, there was little or no slag–metal mixing, hence the rate of desulphurisation was slow. The slag-metal mixing was achieved at high gas flow rates corresponding to $\dot{\varepsilon}$ > 80 W t$^{-1}$ for which the rate constant is

$$k(\text{min}^{-1}) \approx 8 \times 10^{-6}(\dot{\varepsilon})^{2.1} \qquad (9.7)$$

Details on the rate of steel desulphurisation in the ladle will be given later when discussing steel refining in the ladle.

## 9.3 LADLE SLAGS

The ladle slag used in the treatment of the aluminium-killed steel is usually the lime (magnesia)-saturated aluminosilicate, preferably low in iron oxide and manganese oxide contents. Prior to the advent of the ladle furnace, the ladle slag was formed by adding to the tap stream lime-based fluxes, as for example (i) 80 percent lime plus 20 percent spar, (ii) prefused calcium aluminate, or (iii) 50 percent lime plus 50 percent prefused calcium aluminate. The amount of flux added is in the range 5 to 10 kg per tonne of steel, depending on the steel-refining practice.

It should be emphasised here that these solid fluxes should not be added into an empty ladle prior to furnace tapping. The reason being that some of the flux remains on the bottom of the ladle and does not mix with the liquid steel stream. Any material stuck on the ladle bottom may ultimately become unstuck and shoot to the surface of the melt during subsequent steel processing, either at the ladle furnace or at the caster, causing melt eruption resulting in molten steel and slag splashing which can be hazardous to the ladle operators.

It should be noted that the practice of steel desulphurisation developed by the steel industry in the 1970s was based primarily on the concept of pouring liquid steel onto a mixture of granular desulphurising fluxes which were dumped into a preheated ladle shortly before furnace tapping, as described for example in the papers by Hutnik *et al.*[8a], Hawkins[8b] and Mikulecky[8c]. However, in this desulphurising practice, attention was paid to the steel temperature and to the preheated ladle temperature to ensure they were sufficiently high to avoid any skull formation on the bottom or side walls of the ladle, hence to prevent the occurrence of a subsequent melt eruption in the ladle. With this point in view, the tap temperature of steel in the furnace was raised to about 1700 to 1750°C, depending on the practice and steel grade, and the hot face temperature of the ladle lining raised to a high enough temperature at the preheater so that it would be about 1125 to 1175°C at the time of furnace tapping.

In most practices using the ladle furnace, the steel is often tapped with little or no aluminium addition to the ladle. After skimming off the furnace slag from the ladle, the melt is deoxidised with bar aluminium followed by the addition of lime-rich fluxes which form the ladle slag during arc reheating. The so called 'slag conditioner' containing about 68% $CaC_2$, 11% $SiO_2$, 9% $Al_2O_3$, 3% CaO and 4% $CO_2$, is often added at the rate of about 1 kg per tonne of steel to maintain a reducing condition in the slag and also to generate a foamy slag for submerged arc heating in the ladle furnace. The average slag composition after arc reheating is in mass percent: 50–56 CaO, 7–9 MgO, 6–12 $SiO_2$, 20–25 $Al_2O_3$ 1–2 (FeO + MnO), 0.3–0.5 $TiO_2$ and small amounts of S and P.

In the silicomanganese-killed coarse-grain steel of high drawability, as for instance the tyre-cord-grade high-carbon steel, there is a narrow critical range for the total aluminium content. It should be within 15 to 25 ppm to ensure that the deoxidation product is molten aluminosilicate of about spessartite composition ($3MnO \cdot Al_2O_3 \cdot 3SiO_2$), which has a high ductility required in the cold drawing of steel to a thin wire of about 0.2 mm diameter. For this grade of steel, the ladle slag to be used is calcium silicate of wollastonite composition ($CaO \cdot SiO_2$).

## 9.4 CHILL FACTORS FOR LADLE ADDITIONS

The ladle additions of ferro alloys and fluxes will lower the steel temperature in the tap ladle. For the quantity of heat extracted from the steel bath, $\Delta H$ $kJt^{-1}$ steel, the decrease in the bath temperature will be

$$- \Delta T \ (°C) = \ \frac{\Delta H}{790}$$

where 790 $kJt^{-1}°C^{-1}$ is the heat capacity of the liquid steel.

The following are the chill factors for 1 kg alloying element addition per tonne of steel at 1630°C.

| Element : | C | Cr | Fe | Mn | Si | Al |
|---|---|---|---|---|---|---|
| $\Delta T$, °C  : | −5.85 | −1.33 | −1.79 | −1.90 | +1.78 | +0.14 |

For Al-killed steels, the heat of reaction in deoxidation of the steel must be included in calculating the effect of Al addition to the tap ladle on the steel temperature.

$$2Al(RT) + 3[O](1630°C) \rightarrow Al_2O_3(1630°C)$$

In the deoxidation of steel containing 600 ppm O, the heat generated is

−14,830 kJt$^{-1}$ which gives a temperature benefit of $\Delta T = 19°C$. That is, with Al deoxidation of the steel in the tap ladle, the decrease in the steel temperature during furnace tapping will be less by 19°C.

The chill factors for a few ferro alloys are given below.

| Added material for 1% alloying element at 100% recovery | Decrease in steel temperature in the tap ladle, −Δ°C |
|---|---|
| Coke | 65 |
| HC – Fe/50% Cr | 41 |
| LC – Fe/70% Cr | 24 |
| HC – Fe/Mn | 30 |
| Fe/50% Si | 0 |

For 1 kg flux addition per tonne of steel at 1630°C, the chill factors are

| Flux | −ΔT, °C |
|---|---|
| $SiO_2$ | 2.59 |
| $CaF_2$ | 3.37 |
| $CaO$ | 2.16 |
| $CaCO_3$ | 3.47 |
| $CaO \cdot Al_2O_3$ | 2.39 |

Apart from the chill effects of the ladle additions, there are heat losses to the ladle refractory lining and heat losses by radiation during furnace tapping. For the case of aluminium killed steel with the ladle addition of 10 kg (lime + pre-fused calcium aluminate) per tonne of steel, the decrease in steel temperature during furnace tapping to a pre-heated ladle is in the range 55 to 75°C. The heat loss due to the flux addition is nearly balanced by the heat generated in aluminium deoxidation of the steel. Therefore, in the particular tapping practice considered, the heat losses are almost entirely by radiation and conduction into the ladle lining.

## 9.5 REACTIONS OCCURRING DURING FURNACE TAPPING

### 9.5.1 NITROGEN PICKUP

During furnace tapping, there is a momentary entrainment of air bubbles at the place of entry of the tap stream to the steel bath in the ladle. This is demonstrated by the photographs in Fig. 9.10, showing increase in air bubble entrainment with increasing free fall height of a water stream.

From a theoretical analysis[9] for the simplified limiting case of an unbroken steel stream from the furnace to the ladle, the following estimate is made of the volume of transitory air entrainment during furnace tapping at the rate of 30 tonne per minute

$$V \text{ (Nm}^3) = 5.3t - 0.34t^2 \tag{9.8}$$

where $t$ is the duration of furnace tapping in minutes. For a 220 $t$ heat, the transitory air entrainment during furnace tapping is estimated to be about 20 Nm$^3$. With respect to the mass of steel (220 t), this volume of transitory entrainment of air bubbles corresponds to 92 ppm N and 25 ppm O.

Almost all the oxygen in the entrained air bubbles will be picked up by the deoxidants added to the ladle. On the other hand, the extent of nitrogen pickup by the steel from the momentarily entrained air bubbles will be controlled by the rate of the chemical reaction

$$N_2(g) \rightarrow 2[N] \tag{9.9}$$

As discussed in section 2.1.1, oxygen and sulphur in steel retard the kinetics of nitrogen reaction with steel. The higher the concentration of oxygen and/or sulphur, the smaller the amount of nitrogen pickup. The plant analytical data within the hatched area in Fig. 9.11 show nitrogen pickup during tapping of 220 t heats, as affected by the extent of steel deoxidation in the tap ladle; in these heats the steel contained about 0.01% S. There is also the possibility of nitrogen pickup from the ladle additions of ferro alloys which contain some nitrogen.

Fig. 9.10   Photographs showing increase in air bubble entrainment with increasing free fall height of a water stream.

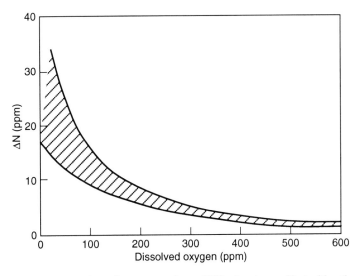

Fig. 9.11   Nitrogen pickup during tapping of 220 t heats as affected by the extent of deoxidation in the tap ladle.

### 9.5.2 HYDROGEN PICKUP

The moisture contained in the ladle additions, e.g. burnt lime, coke and ferro alloys, will react with liquid steel resulting in hydrogen pickup by the steel during furnace tapping.

$$H_2O \rightarrow 2[H] + [O] \tag{9.10}$$

Noting that the solubility of hydrogen in liquid steel increases with decreasing oxygen content, the hydrogen pickup in the aluminium-killed steel is expected to be somewhat higher.

Of the ferro alloys added, the ferromanganese is probably the major contributor to the hydrogen pickup. Plant studies made on changes in hydrogen levels at various stages of steelmaking processes indicate that an average hydrogen pickup due to ferromanganese addition is $\Delta H = 2$ ppm H/%Mn.

### 9.5.3 FURNACE SLAG CARRY OVER

The percentages of utilisation of the ladle additions for low carbon heats are 85–95% for Mn, 60–70% for Si, and 35–65% for Al. The copious black fumes emitted at the time of the addition of ferromanganese to the tap stream, indicate that the vaporisation of manganese is the primary cause of some loss of manganese during the ladle addition. Losses in the ladle additions of aluminium and silicon are due to reactions with the furnace slag that is carried into the tap ladle. There is also the iron oxide rich skull accumulating at the converter mouth, some of which falls into the ladle during furnace tapping and reacts with aluminium and silicon.

*Mass of slag carryover and fallen converter skull:*

The reaction of aluminium and silicon with the furnace slag and fallen converter skull is represented in a general form by the following equation

$$Fe(Mn)O_x + Al(Si) \rightarrow Fe(Mn) + Al(Si)O_x \tag{9.11}$$

Using the average molecular masses and assuming 80% $Fe_3O_4$ for the converter skull, the following approximate relation is derived for the loss of aluminium and silicon to the ladle slag for 220 t steel in the tap ladle.

$$[\%Al + \%Si]_s = \Delta(\%FeO + \%MnO)W_{fs} \times 10^{-6} + W_{sk} \times 10^{-4} \tag{9.12}$$

where

$W_{fs}$ = mass of furnace slag carryover, kg
$W_{sk}$ = mass of fallen converter skull, kg
$\Delta(\%FeO_t + \%MnO)$ = decrease in oxide contents of furnace slag during tapping.

For low carbon Q–BOP heats, $\Delta (\%FeO_t + \%MnO)$ is about 20% for which the above equation becomes

$$[\%Al + \%Si]_s = (0.2W_{fs} + W_{sk}) \times 10^{-4} \tag{9.13}$$

*Phosphorus reversion:*

Another consequence of slag carryover is phosphorus reversion to the Al-killed steel in the tap ladle. For a 220 t heat, the phosphorus reversion $\Delta[ppm\ P]$ is represented by

$$\Delta[ppm\ P] = 0.045\Delta(\%P)W_{fs} \tag{9.14}$$

where $\Delta(\%P)$ is the decrease in the phosphorus content of the furnace slag carried into the tap ladle. For low carbon Q–BOP heats, $\Delta(\%P)$ is about 0.3% for which the above equation becomes

$$\Delta[ppm\ P] \approx 0.014W_{fs} \tag{9.15}$$

Results of the analysis of Q–BOP heat logs, as described above, are plotted in Fig. 9.12 (the data points are scattered within the hatched area) which show the expected increase in the extent of phosphorus reversion with an increase in $[\%Al + \%Si]_s$ reacting with the ladle slag. For high phosphorus reversion of 45 ± 10 ppm, the mass of slag carryover is estimated to be $W_{fs}$ = 3214 ± 714 kg, corresponding to a slag thickness (in a 220 t ladle) of about 125 ± 28 mm. In most cases of furnace tapping, the phosophorus reversion is about 20 ± 10 ppm for which the estimated slag carryover is about 1430 ± 714 kg. This estimated average slag carryover during furnace tapping corresponds to a slag thickness (in a 220 t ladle) of

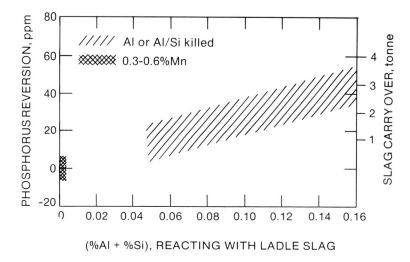

Fig. 9.12   Phosphorus reversion during tapping of 220 t Q–BOP heats is related to %Al + %Si reacting with carry-over furnace slag and fallen converter skull.

about 55 ± 28 mm, which is in general accord with the plant observations in the so-called minimum slag carryover practice.

When the heat is tapped open, with only the addition of ferromanganese and a small amount of aluminium, the steel is not reduced enough to cause phosphorus reversion. In fact, in some cases of open heat tapping with about 0.3–0.6% Mn addition, the phosphorus content of the steel decreases by about 10 ppm, due to the mixing of the carried over furnace slag with the steel during tapping.

For the average value of $W_{fs}$ = 1430 ± 714 kg, for which Δ[ppm P] = 20 ± 10, $W_{sk}$ is estimated to be 470 ± 35 kg.

## 9.6 LADLE FURNACE

The ever increasing users demand for higher quality steel, particularly for critical applications, and the advent of sequential continuous casting, made it necessary for steel plants to install ladle furnaces for steel reheating that is needed for (i) the extended time of steel refining and (ii) the adjustment of steel temperature for the caster on a correct timely basis.

There are several types of ladle furnaces for arc reheating of steel, designed by different manufacturers, e.g. Finkl-Mohr, Lectromelt, ASEA–SKF, MAN–GHH, Daido–NKK etc. The basic features of the Daido-NKK ladle furnace are shown in Fig. 9.13.

### 9.6.1 ARC REHEATING IN THE LADLE FURNACE (LF)

The following statements summarise the basic requirements for efficient arc reheating in the ladle furnace.

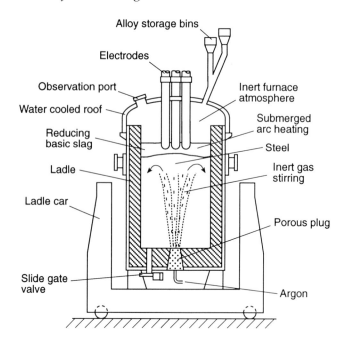

Fig. 9.13   Sketch of Daido–NKK ladle furnace.

A. To shorten reheating time, minimise refractory erosion and increase the efficiency of energy consumption, apply the following:
   (i)    use a large capacity transformer, e.g. 35 to 40 MW for a 200 to 250 tonne heat
   (ii)   practice submerged arcing in the slag layer
   (iii)  apply argon stirring with bottom porous plug at a flow rate of about 0.5 Nm³ min⁻¹.
B. For submerged arc heating, reduce the arc length by a large current and low voltage operation at the LF.
C. For high heating efficiency with a stable submerged arc and low refractory erosion, the thickness of the slag layer should be about 1.3 × arc length. The calculated arc length and electric power of arc are shown in Fig. 9.14 as functions of voltage and secondary current.
D. To ensure submerged arc heating, the tap voltage and secondary current are set in accordance with the slag thickness, before the start of arcing.
E. With the tap voltage of 385 to 435 V and secondary current 40 to 50 kA, the liquid steel temperature is raised at the rate of 3.5 to 4.5°C min⁻¹.
F. Energy Efficiency and Electrode Consumption:
   The energy efficiency $E$ of heating is defined by the ratio

$$E = \frac{\text{Temperature increase, } \Delta°C}{\text{Energy consumption, kWh t}^{-1}} \times 0.22$$

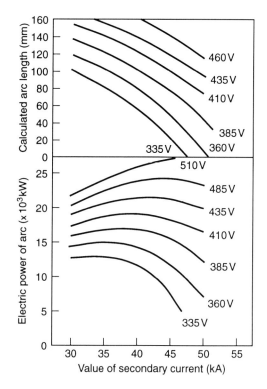

Fig. 9.14   Calculated arc lengths and electric power of arc.

where 0.22 kWh $t^{-1}$ $°C^{-1}$ is the theoretical amount for 100 percent efficiency of energy consumption.

The energy efficiency of heating increases with an increasing heat size.

| Heat size, t | E |
|---|---|
| 50 | 0.40 to 0.50 |
| 100 | 0.45 to 0.55 |
| 250 | 0.55 to 0.65 |

G. The electrode consumption increases with an increasing
(i)   cross sectional current density, i.e. with a decreasing electrode diameter;
(ii)  heating time.

| Electrode consumption kg $t^{-1}$ | Current density A $cm^{-2}$ | Heating time min |
|---|---|---|
| 0.1 to 0.3 | 10 | 15 |
| 0.4 to 0.8 | 40 | 80 |

H. The arcing periods are usually limited to less than 15 minutes at one time of arcing.

9.6.2 REHEATING WITH OXYGEN INJECTION

Liquid steel can be reheated by oxidising Al and Si with a lance injection of oxygen deep into the melt. For the reactions

$$2Al(R.T.) + \tfrac{3}{2}O_2(RT) \rightarrow Al_2O_3(1630°C) \qquad (9.16)$$

$$Si(R.T.) + O_2(RT) \rightarrow SiO_2(1630°C) \qquad (9.17)$$

heats generated are

27,000 kJ/kg Al
28,500 kJ/kg Si in Fe–75% Si alloy

To raise the melt temperature by 35°C with 70% thermal efficiency of the process, the amounts of reactants used are

1.46 kg Al/t steel and 1 Nm³O₂/t steel
1.85 kg Fe–75% Si/t steel and 1.2 Nm³O₂/t steel

Reheating with Al oxidation in the melt is being used in RH–OB degassers and in various types of ladle furnaces, including the Bethlehem plants[12] in Steelton, Sparrows Point and Burns Harbor. With this practice, the steel temperature can be raised at the rate of 6°C min⁻¹. To ensure homogenisation of the melt temperature and flotation of oxidation products out of the melt, the oxygen injection must be followed by argon stirring and rinsing.

During oxygen injection to oxidise Al, there is also some oxidation of Si, Mn and C as shown below:

i.   In all grades of steel, 0.6 g Si and Mn/Nl O₂
ii.  In C ranges < 0.4%, a little or no C oxidation
iii. In rail grades (≈0.8% C), the loss is about 0.01% C per heat.


## 9.7 STEEL DEOXIDATION

For liquid steel to be castable, hot workable and the product to have the desired metallurgical and mechanical properties, the liquid steel must be deoxidised in the tap ladle for which the deoxidants used are: ferromanganese, ferrosilicon, silicomanganese and aluminium.

There are three categories of steel deoxidation:

(i)   Resulphurised steel deoxidised with ferromanganese to 100–200 ppm O.
(iia) Semi-killed steel deoxidised with Si/Mn to residual 50 to 70 ppm O.
   b) Semi-killed steel deoxidised with Si/Mn/Al to residual 40 to 25 ppm O.

c) Semi-killed steel deoxidised with Si/Mn/Ca to residual 20 to 15 ppm O.

(iii)  Killed steel deoxidised with Al to residual 4 to 2 ppm O.

The reaction equilibrium data for steel deoxidation were presented in section 6.8. In this section the practical aspects of deoxidation are presented with reference to the test results obtained at the U.S. Steel plants.

## 9.7.1 OPEN HEAT TAPPING

Now that the ladle furnace is available for the final refining of steel, the steel can be partially deoxidised in the tap ladle with Fe/Mn and Fe/Si, followed by a final deoxidation with aluminium at the ladle furnace station. There are several advantages to this practice, as for example

i.   to minimise nitrogen pickup during tapping
ii.  to minimise phosphorus reversion from carried over furnace slag
iii. to eliminate aluminium losses in the tap ladle.

## 9.7.2 SLAG AIDED DEOXIDATION

The dissolution of the deoxidation products Mn(Fe)O or manganese silicates in a neutral ladle slag, such as calcium aluminate, will lower the thermodynamic activities of the deoxidation products, hence increase the extent of deoxidation. The concept of slag aided deoxidation is by no means new. In the Perrin process developed in the early 1930s, the deoxidation of the open hearth or Bessemer steel with ferromanganese and ferrosilicon was enhanced by tapping the steel into a molten calcium (magnesium) aluminosilicate slag placed on the bottom of the tap ladle.

### 9.7.2a Partial deoxidation with Fe/Mn

The plot in Fig. 9.15 depicts the progress of partial deoxidation of steel during tapping of a 200 t heat with the addition of 1800 kg lime-saturated calcium aluminate and ferromanganese to the ladle in the early stage of tapping, e.g. at 1/8 ladle fillage. At the time of ladle additions, the small quantity of steel in the ladle is almost completely deoxidised with the residual manganese in the steel being at a concentration of about 1.6% Mn. As the ladle is filled, the dissolved manganese is consumed in the deoxidation reaction and diluted to about 0.32% Mn when the ladle is full and the residual dissolved oxygen is reduced to about 300 from 650 ppm in the furnace.

The results obtained using this practice in the EAF and Q–BOP shops of U.S. Steel are plotted in Fig. 9.16 for steels containing in the tap ladle about 0.003% Si and Al each. If no calcium aluminate slag addition were made to the tap ladle, the deoxidation by Mn and Fe alone with pure Mn(Fe)O as

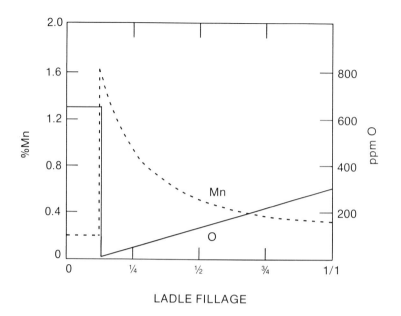

Fig. 9.15   Changes in dissolved contents of Mn and O in steel during furnace tapping with calcium aluminate and ferromanganese added to the tap ladle at 1/8 ladle fillage. From Ref. 10.

the deoxidation product would have resulted in much higher levels of the residual dissolved oxygen in the steel, as depicted by the dotted curve.

A more detailed analysis of the EAF heats is given in Fig. 9.17. The slopes of the lines $K'_{FeMn}$ are in close agreement with the slag–metal equilibrium relation given in Fig. 6.2. Similarly, the relation between %MnO in the ladle slag and the product [%Mn][ppm O] in Fig. 9.18 is in close accord with the equilibrium value.

In these EAF trial heats, there was no argon stirring in the ladle during furnace tapping. Yet, the slag aided partial deoxidation achieved during furnace tapping was close to the levels governed by the slag–metal equilibrium. Evidently there is sufficient mixing of slag and metal for the deoxidation reaction to progress at a relatively fast rate. Because of the off-centre entry of the steel stream into the ladle, there is always a slow rotation of the slag layer in the ladle. Thus, most of the slag layer passes through the area of stream entry where there is localised mixing of slag and metal, hence a faster reaction rate.

In this practice, the final aluminium deoxidation was done with a wire feeder, based on the level of the residual dissolved oxygen in the tap ladle. The efficiency of aluminium recovery was about 85 to 90 percent. Although the ladle slag contained 4 to 7% MnO and 7 to 12% FeO, there was little reduction of these oxides during the injection of aluminium wire, accompanied by argon stirring at a moderate rate of about 200 Nl min⁻¹.

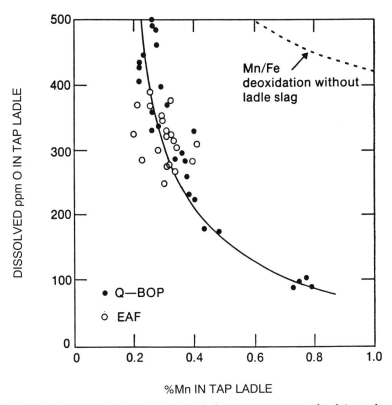

Fig. 9.16 Partial deoxidation of steel with ferromanganese and calcium aluminate ladle slag during furnace tapping for steel in ladle containing Si and Al each < 0.003%. From Ref. 10.

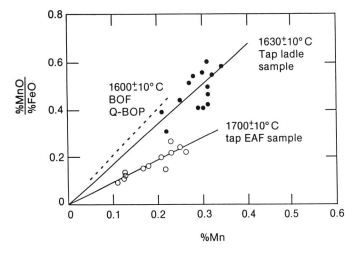

Fig. 9.17 Manganese content of steel in EAF at tap and in tap ladle is related to %MnO/%FeO ratio in furnace and ladle slags. From Ref. 10.

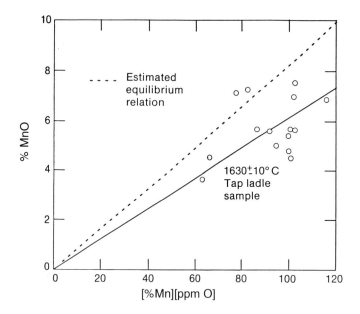

Fig. 9.18 Manganese deoxidation product is related to MnO content of ladle slag after EAF tapping. From Ref. 10.

In the EAF steelmaking, the nitrogen content of steel in the furnace is 35 to 45 ppm at tap. With partial deoxidation from about 650 to 300 ppm in the tap ladle, the nitrogen pickup will be less than 10 ppm.

### 9.7.2b Deoxidation with Si/Mn

In semi-killed steels deoxidised with silicomanganese and a small amount of aluminium, the deoxidation product is molten manganese aluminosilicate and the residual dissolved oxygen is about 50 ppm for steel containing about 0.8% Mn and 0.2% Si. With the addition of 1000 kg of prefused calcium aluminate to the ladle for a 200 t heat, the residual dissolved oxygen in the steel can be lowered to about 20 ppm with the Si/Mn deoxidation; this is because of the dissolution of the deoxidation products in the calcium aluminate slag.

It should be noted that with ladle slag containing a high percentage of alumina and a low concentration of silica, some alumina will be reduced by silicon. As noted from the equilibrium data in Fig. 9.19, steel containing, for example, 0.2% Si may pick up an appreciable amount of aluminium by the reduction of alumina from the slag. The final stage of deoxidation of the steel bath in the ladle is controlled by the aluminium recovered from the slag.

### 9.7.2c Deoxidation with Al

The oxygen sensor readings and analysis of steel samples from the tap ladle have indicated that the addition of lime+spar or prefused calcium

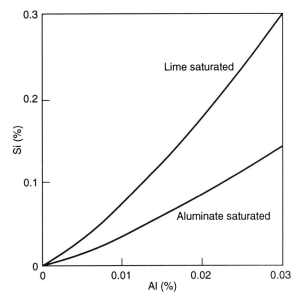

Fig. 9.19   Si and Al contents of steel in equilibrium with molten calcium aluminate slags containing $\approx 5\%$ $SiO_2$ at 1600°C.

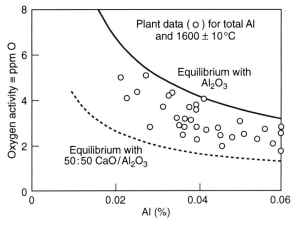

Fig. 9.20   Plant data for deoxidation with Al and calcium aluminate ladle slag during furnace tapping compared with equilibrium values for $Al_2O_3$ and molten calcium aluminate (50:50 CaO/$Al_2O_3$) inclusions.

aluminate to the ladle extends the deoxidation of steel with aluminium during furnace tapping. This is demonstrated by the plant data plotted in Fig. 9.20. The oxygen sensor readings give results which are between those for the deoxidation product $Al_2O_3$ and molten calcium aluminate, indicating that the alumina inclusions in the melt are in fact fluxed by the ladle slag.

In the plant trials with 30 t AOD heats, reported by Riley and Nusselt[11], the oxygen sensor readings were taken after vigorous mixing of steel with the

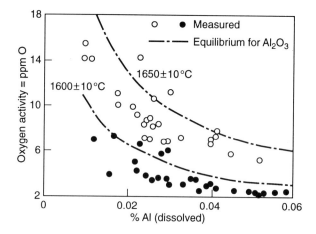

Fig. 9.21   Oxygen sensor measurements after Al deoxidation and hard argon stirring in 30 t AOD, reported by Riley and Nusselt.[11]

lime-rich aluminate slag and hard argon blowing. As can be seen from their plant data in Fig. 9.21, the residual dissolved oxygen contents for 1650 and 1600°C are below the equilibrium curves for pure alumina, because of the fluxing of alumina inclusions with the calcium aluminate furnace slag and argon stirring. In fact, the oxide inclusions separated from the steel samples were found to have compositions similar to those of the calcium aluminate AOD slag.

### 9.7.3 DEOXIDATION WITH CA–SI

Steel semi-killed with Si/Mn can be deoxidised further by the cored-wire injection of Cal-Sil. In the 60-tonne trial heats made at the Anzin Works in France, reported by Thomas *et al.*[12] the semi-killed steel (0.1% C, 0.2% Si, 0.5% Mn) was deoxidised from 75 to about 15 ppm O by the cored-wire injection of Cal–Sil (30% Ca). Their reported plant data are plotted in Fig. 9.22 showing the fractions of oxygen and calcium reacting as a function of the addition of Ca.

   If the content of dissolved oxygen in the ladle is lowered to 20 or 30 ppm by slag-aided Si/Mn deoxidation during furnace tapping, a subsequent deoxidation by Cal–Sil injection may lower the residual dissolved oxygen to about 10 ppm or less.

### 9.7.4 ARGON RINSING FOR INCLUSION FLOATATION

After arc reheating and trim additions for composition adjustment, the rate of argon bubbling should be reduced to a low level of 75 to 150- Nl min$^{-1}$, i.e. argon rinsing, to float out the residual deoxidation products, known as inclusions. With argon rinsing for 5 to 8 min, the total oxygen content of the steel as simple or complex oxide inclusions is less than 15 ppm.

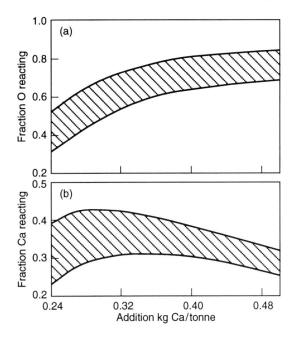

Fig. 9.22 Fractions of oxygen and calcium reacting in deoxidation of semi-killed steel (~ 75 ppm O) with Cal–Sil (30% Ca) cored-wire injection are related to the amount of Ca added, derived from plant data reported by Thomas *et al.*[12]

At argon flow rates above 200 Nl min$^{-1}$, the gas bubbles open up an 'eye' of 30 to 60 cm in the slag layer, thus exposing the steel surface to air. This situation leads to the re-oxidation of the bath, hence generates more oxide inclusions and lowers the efficiency of inclusion removal.

The total aluminium content of the metal sample taken from the tundish or the mould is about 0.007% lower than the content in the last sample taken from the ladle. This fade in aluminium is due primarily to the reoxidation of steel during casting and to the flotation of alumina out of the melt in the tundish.

## 9.8 DESULPHURISATION

In certain grades of clean steel, e.g. armour plates, plates for off-shore oil installations and line pipes, the sulphur content should be less than 20 ppm. This low level of sulphur in the steel can be achieved only by desulphurisation of the fully killed steel in the ladle. The ladle slag to be used is the lime-saturated calcium aluminate containing less than 10% $SiO_2$ and less than 0.5% of both FeO and MnO.

The overall desulphurisation reaction is represented by

$$(CaO) + [S] + \tfrac{2}{3}[Al] \rightarrow (CaS) + \tfrac{1}{3}(Al_2O_3) \tag{9.18}$$

For this reaction to proceed rapidly, good slag-metal mixing is necessary; this is achieved by top lance injection of argon at the rate of 1.0–1.8 Nm³ min⁻¹. It should be noted that with the porous plug on the ladle bottom, the maximum argon flow attainable is about 0.6 Nm³ min⁻¹. In some steel works, three porous plugs are installed on the bottom of the ladle for hard argon stirring. However, in most plants hard argon stirring for de-sulphurisation is done by top lancing deep into the melt in the ladle.

To prevent re-oxidation of the steel bath and nitrogen pick-up, the ladle must be covered with a lid; the free board under the lid being flushed with argon at the rate of about 0.6 Nl min⁻¹. Furthermore, there should be at least 75 cm deep free board from the ladle top to the slag surface to prevent splashing of slag and metal on to the ladle lid. Many ladle furnaces do not have high enough free board for extensive steel de-sulphurisation with heavy argon stirring.

### 9.8.1 DESULPHURISATION METHODS

#### 1. TN (Thyssen–Niederrheim) process:

The ladle slag is made up by adding to the tap ladle, per tonne of steel, about 8 kg of the mixture 80% CaO + 20% CaF₂. The alumina generated in the deoxidation of the steel during furnace tapping reacts with the lime and spar added to the ladle which forms a lime-rich molten aluminate. The aluminium content of the steel is adjusted to 0.07–0.10% and the steel is desulphurised by top lance injection of either Cal–Sil (30% Ca) or a powder mixture of CaO and CaF₂, with argon flowing at the rate of about 1 Nm³ min⁻¹. For grades of steel low in silicon content, the desulphurisation is done by a lime + spar injection. Depending on the amount injected, 50 to 80 percent desulphurisation is achieved in 12 to 15 minutes time of treatment.

| Amount injected kg t⁻¹ | Percent desulphurisation Cal–Sil | Percent desulphurisation CaO–CaF₂ |
|---|---|---|
| 1 | 55 | 45 |
| 2 | 70 | 60 |
| 3 | 80 | 70 |

With this method, the steel can be desulphurised to a level of 20 or 40 ppm S, provided the initial sulphur content is less than 110 ppm $S_o$.

In a modified version of the TN process, the steel is desulphurised with a co-injection of Cal–Sil (30% Ca) and a powder mixture of 7% Al + 15% CaF₂ + 78% CaO deep into the melt. The following quantities are used per tonne of steel:

(i)   10 to 15 kg calcium aluminate-based ladle slag with CaO/Al₂O₃ = 1.5 to 1.8;
(ii)  0.8 to 1.3 kg Cal–Sil injected at the rate of 35 kg min⁻¹;
(iii) 1 to 2 kg (Al–CaF₂–CaO) injected at the rate of 60 kg min⁻¹.

Using this method, the steel initially containing 0.014 ± 0.002% $S_o$ is de-sulphurised to 25 to 45 ppm S in 6 or 7 minutes time of co-injection.

*2. Calcium aluminate slag and argon stirring:*

If there is sufficient free board over the slag layer in the ladle, ≈ 1 m, extensive desulphurisation can be achieved with a lime-saturated alumi-nate slag and hard argon stirring at the rate of 1.5 to 1.8 $Nm^3$ $min^{-1}$. This practice is best suited for operating in the tank degasser, where there is a very good slag–metal mixing at reduced tank pressures. Furthermore, there is hydrogen and some nitrogen removal during desulphurisation.

The slag/metal sulphur distribution ratios at the end of desulphurisa-tion are plotted in Fig. 9.23. The plant data from U.S. Steel Works are within the hatched area for desulphurisation of various grades of steel by (i) TN process, (ii) with lime-rich aluminate slag and hard argon stirring in a covered ladle and (iii) in the tank degasser. The sulphur distribution ratios after desulphurisation are in close agreement with the slag–metal equilibrium values for 0.006 ± 0.002 and 0.03 ± 0.01% Al at temperatures of 1600 ± 15°C. The lower the silica content of the aluminate slag the greater the extent of desulphurisation.

As reported by Bannenberg *et al.*[13] of Dillinger Hüttenwerke, high sul-phur steels can be desulphurised to low residual levels by hard argon stirring in the tank degasser, using the lime-rich aluminate slag containing less than 1% $SiO_2$. In this practice with 180 tonne heats, burnt lime and bar aluminium are added to the tap ladle and the furnace slag carry over is

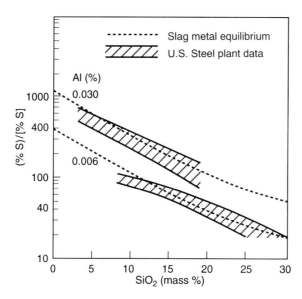

Fig. 9.23 Slag/metal sulphur distribution ratios after desulphurisation with lime-saturated CaO–MgO–$Al_2O_3$–$SiO_2$ slags at 1600 ± 15°C with steels containing residual 0.006 ± 0.002% Al or 0.03 ± 0.01% Al are compared with the equilibrium values.

kept to about 400 kg. The initial and final sulphur contents and quantities of slag in the ladle are as shown below.

| Ladle slag kg t$^{-1}$ | Sulphur, ppm Initial | Final |
|---|---|---|
| 22 | 700 | 60 |
| 16 | 500 | 40 |
| 14 | 300 | 20 |
| 12 | 200 | 15 |
| 10 | 100 | 10 |

If the silica content of the slag is high and the silicon content of steel is low, there will be a reduction of silica from the slag by aluminium in the steel during hard argon stirring.

$$SiO_2 + \tfrac{4}{3}[Al] \rightarrow [Si] + \tfrac{2}{3}(Al_2O_3) \tag{9.19}$$

It is seen from the plot in Fig. 9.24 that the tank degasser data[13] are consistent with the experimental equilibrium data[14] for lime-saturated calcium aluminate melts containing less than 5% $SiO_2$.

### 3. EXOSLAG practice:

The Al-killed steel can be desulphurised, at least partially, during furnace tapping by pouring steel onto a molten lime-saturated calcium aluminate, formed prior to tapping by the exothermic alumino-thermic reaction.

$$Fe_2O_3 + 2Al \rightarrow Al_2O_3 + 2Fe \tag{9.20a}$$

$$xCaO + Al_2O_3 \rightarrow xCaO \cdot Al_2O_3 \tag{9.20b}$$

An exothermic slag mixture, EXOSLAG, was formulated and tested at the Research Laboratory of USS in 1984.[15] The following preferred mixture includes the additional lime needed to flux the alumina generated in the deoxidation reaction to give a lime-saturated aluminate slag with the nominal ratio of $\%CaO/\%Al_2O_3 = 60/40$.

Mixture must be kept dry
- 58% burnt lime
- 30% haematite ore fines (low $SiO_2$)
- 12% aluminium powder

The ignition temperature of the exothermic reaction in equation 9.20a for this mixture was determined in laboratory tests to be in the range 870 to 890°C which is 130 to 110°C lower than the hot-face temperature of the lining in the preheated ladle.

In a preheated ladle, with the hot-face temperature of about 1000°C, sufficient self ignition of the EXOSLAG occurs in about 5 minutes to permit tapping of the steel onto the desired molten calcium aluminate slag. The self ignited material is only partly molten because of the presence of excess lime, which is fluxed subsequently with the alumina generated in the deoxidation of steel with Al.

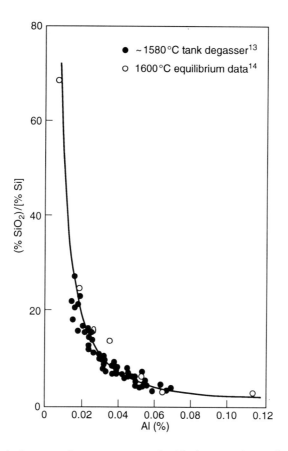

Fig. 9.24   Tank degasser data are compared with the experimental equilibrium data for aluminium reduction of silica from lime-saturated calcium aluminate melts containing $SiO_2 < 5\%$.

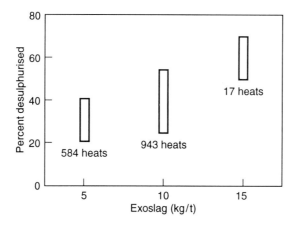

Fig. 9.25   Percent sulphur removed during furnace tapping increasing with an increase in amount of EXOSLAG in tap ladle. From Ref. 15.

The extent of desulphurisation achieved with the EXOSLAG practice for 200-tonne EAF, BOF and Q–BOP heats in USS are shown in Fig. 9.25 for tap steel compositions in the range 0.04 to 0.4% C and 0.006 to 0.045% S, with the amount of furnace slag carry over < 1500 kg.

There is also the temperature benefit realised in the EXOSLAG practice. As indicated by the plant data in Fig. 9.26 the tap temperature can be lowered by about 24°C with the EXOSLAG practice. In this practice there is 20 to 30 ppm N pickup by the steel.

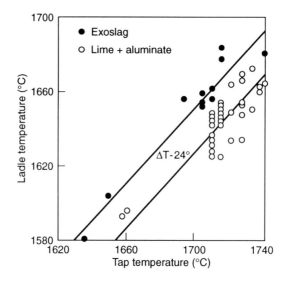

Fig. 9.26   Decrease in steel temperature from furnace to ladle. From Ref. 15.

It should be emphasized that the EXOSLAG mixture put on the bottom of the preheated ladle must be *dry* to prevent violent eruption when the liquid steel is poured upon it. Also, as soon as the exothermic reaction has completed forming molten calcium aluminate, with excess lime floating on top, the steel must be tapped right away. Molten slag formed on the bottom of the ladle should not be allowed to solidify before the start of furnace tapping. Pouring liquid steel onto a mass of solid slag at the bottom of the ladle can have hazardous consequences, caused by subsequent floatation of the mass of slag to the melt surface in the ladle.

9.8.2 RATE OF DESULPHURISATION

The relation between the argon flow rate and the density of power of stirring by injection about three metres below the melt surface is shown in Fig. 9.27, calculated using equation (9.4). As pointed out earlier, the slag–metal mixing is achieved at high gas flow rates corresponding to $\dot{\varepsilon} > 80$

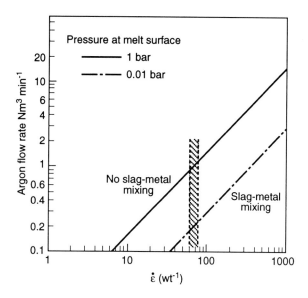

Fig. 9.27   Density of power of stirring by gas injection at 3m below the surface of a 200 t melt.

$W/t^{-1}$. For an argon flow rate of 1.5 $Nm^3$/min at 3m depth of injection, $\dot{\varepsilon}$ = 100 $W/t^{-1}$ for which the rate constant is $k$ = 0.13 $min^{-1}$ for desulphurisation as given in equation 9.21.

$$\ln \frac{\%S - \%S_{eq}}{\%S_o - \%S_{eq}} = -kt \qquad (9.21)$$

For a slag/metal mass ratio of 1:100 in the ladle, the sulphur content of steel in equilibrium with the slag is given by

$$\%S_{eq} = \frac{100[\%S_o]}{100 + \{(\%S)/[\%S]\}_{eq}} \qquad (9.22)$$

where $\{(\%S)/[\%S]\}_{eq}$ is the equilibrium distribution ratio derived from the equilibrium data in Figs. 6.11 and 6.12 and the anticipated %Al in the steel at the end of desulphurisation. As is seen from the calculated rate curves in Fig. 9.28, an increase in the silica content of the slag lowers the extent of desulphurisation because of an increase in $\%S_{eq}$ with an increasing $\%SiO_2$. With $SiO_2$ < 10% and initial $S_o$ < 110 ppm, the liquid steel is desulphurised to 30–20 ppm with the lime-rich aluminate slag and hard argon stirring. Further desulphurisation of steel to levels below 10 ppm is achieved by flux injection. The plant data in Fig. 9.29, reported by Oguchi *et al.*[16], are for the powder mixture (80% CaO + 20% $CaF_2$) injected with argon at the rate of 0.7 kg $min^{-1}t^{-1}$.

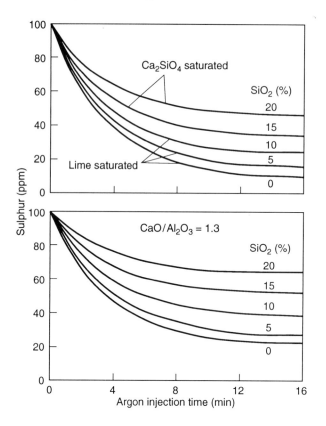

Fig. 9.28 Calculated rate curves for desulphurisation at 1600°C of a 200 t steel containing 0.03% Al with 2 t ladle slag and argon stirring at 1.5 Nm³min⁻¹ (≡ ε̇ 100 Wt⁻¹).

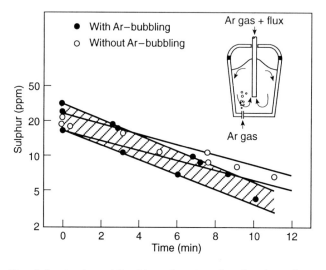

Fig. 9.29 Desulphurisation of liquid steel to very low levels with injection of (80% CaO + 20% CaF₂). From Ref. 16.

## 9.9 VACUUM DEGASSING

Since the 1950s many advances have been made in plant equipment for vacuum degassing of steel and in the technology of steel refining in vacuum degassing facilities. Only brief comments are made here on different types of degassers.

### 9.9.1 VACUUM DEGASSER WITH ARC REHEATING

There are various types of ladle furnaces attached to a vacuum system; only two are briefly mentioned here. In the Finkl–Mohr system, the electrodes for arc reheating are inserted through the top of the vacuum tank cover and the melt is stirred with gas bubbling through a porous plug. The ASEA–SKF system consists of a ladle furnace, a mobile induction coil stand (for induction stirring of the melt), a vacuum cover with an exhaust duct and a cover fitted with three carbon electrodes.

### 9.9.2 VACUUM LADLE DEGASSING

Steel is treated in a tank degasser without arc reheating as shown schematically in Fig. 9.30 for two different melt stirring systems. In one case, the liquid steel is inductively stirred during degassing and alloying additions. In the other case, the melt is stirred with argon bubbling through a porous plug on the ladle bottom. As discussed earlier, extensive steel desulphurisation can be achieved in the tank degasser in addition to hydrogen and some nitrogen removal.

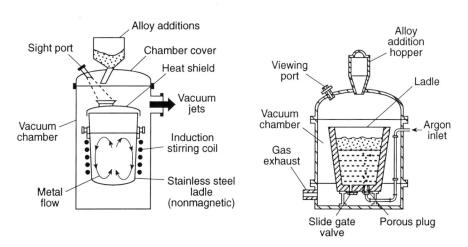

Fig. 9.30   Two types of tank degasser: (a) induction coil stirring, (b) porous plug for argon bubbling.

### 9.9.3 RECIRCULATION DEGASSING

In recirculation degassing, the liquid steel in a ladle is forced by at-mospheric pressure into an evacuated chamber where it is exposed to low pressure and returned back into the ladle. The steel is recirculated through the low pressure chamber 40 to 50 times to achieve the desired level of degassing. There are three types of degassers.

(a) DH (Dortmand–Hoerder) degasser with one snorkel and ladle mov-ing up and down to force the liquid metal in and out of the vacuum chamber.

(b) RH (Ruhrstahl-Heraeus) degasser with two snorkels and stationary ladle (Fig. 9.31a). Argon is injected into one snorkel to force liquid steel into the vacuum chamber; steel flows back into the ladle through the other snorkel.

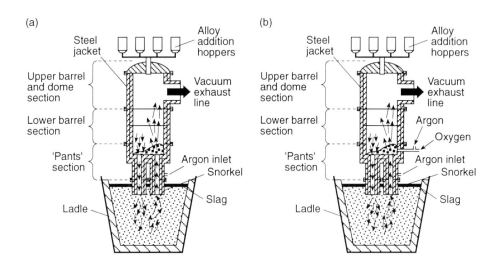

Fig. 9.31   General outline of (a) RH and (b) RH–OB vacuum degassers.

(c) In the RH–OB degasser, developed by Nippon Steel Corp., there is oxygen injection into the pool of liquid steel in the vacuum chamber (Fig. 9.31b). With this degasser it is possible to (i) reheat the steel by oxidising aluminium added to the melt and (ii) enhance steel decarburisation.

In the RH and RH–OB, the rate of steel circulation between the ladle and vacuum chamber is determined by the snorkel diameter and rate of argon injection into the snorkel as shown in Fig. 9.32a. The extent of temperature loss during vacuum degassing becomes greater with a decreasing steel capacity of the unit (Fig. 9.32b). Similar heat losses from the melt will occur in the tank degasser. In the use of these processes, heats to be degassed are tapped at a higher temperature depending on the heat size.

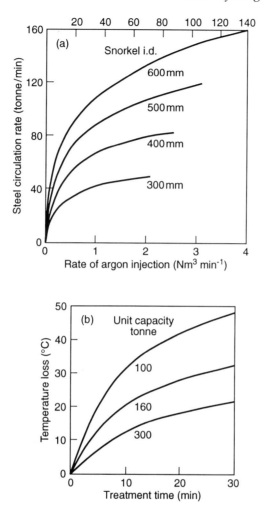

Fig. 9.32   RH and RH–OB plant data for (a) steel circulation rate, (b) temperature loss during treatment time.

### 9.9.4 CARBON DEOXIDATION

One of the applications of RH or RH–OB treatment of steel is to lower the level of oxygen content of steel by carbon deoxidation before adding aluminium for complete deoxidation. For the light vacuum treatment, BOF steel is tapped at about 0.08 to 0.10% C when the oxygen content is 400 to 300 ppm. During degassing, the steel is decarburised down to 0.04% C and deoxidised to about 200 ppm O which is four times less than the oxygen content of BOF steel with 0.04% C at turndown for tap. With this carbon-deoxidation practice there is a considerable cost saving in aluminium usage for deoxidation.

### 9.9.5 DECARBURISATION

The ultra low-carbon steel is made by vacuum degassing at a pressure below 1 mbar. The carbon and dissolved oxygen contents of the steel decrease in the vacuum treatment by the reaction

$$[C] + [O] \rightarrow CO(g) \tag{9.23}$$

The carbon and oxygen contents of steel before and after RH treatment are shown in Fig. 9.33. Similar results are obtained with the tank degasser. Although the pressure in the vacuum chamber is at about 0.001 atm, the final carbon and oxygen contents are closer to the equilibrium values for CO pressures of 0.06 to 0.08 atm. After about 20 minutes of treatment time, the final oxygen content of the steel is always high when the initial content is high, i.e. initial carbon content is low, in both the tank degasser and the RH degasser (Fig. 9.34).

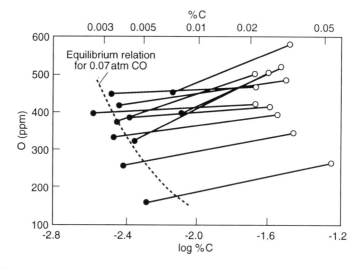

Fig. 9.33   Carbon and oxygen contents of steel before (○) and after (●) R H treatment.

A decrease in the dissolved oxygen content of the steel in vacuum decarburisation is always less than that expected from the stoichiometry of the reaction in equation (9.23).

It is evident that there is oxygen transfer from the ladle slag to steel during vacuum decarburisation. Another source of oxygen is the iron oxide-rich skull buildup on the inner walls of the vacuum chamber from previous operations. Therefore, the other reaction to be considered is

$$(FeO) + [C] \rightarrow Fe + CO(g) \tag{9.24}$$

From the stoichiometry of the reactions, the material balance gives the

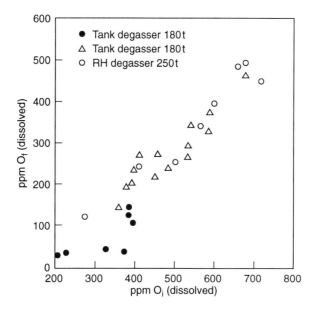

Fig. 9.34   Oxygen contents of steel before and after vacuum decarburisation. Points
● Δ ○. Ref. 13 17 18.

following relation for the amount of oxygen transferred to the steel from
the ladle slag and oxidised skull in the vacuum chamber,

$$\Delta O\ (slag) = {}^{16}\!/_{12}[\Delta C] - [\Delta O] \qquad (9.25)$$

where
$$[\Delta C] = [\%C]_i - [\%C]_f$$
$$[\Delta O] = [ppm\ O]_i - [ppm\ O]_f$$

The values of ΔO (slag) derived from the plant data using the above
relation for vacuum decarburisation are shown in Fig. 9.35. The higher the
initial carbon content, the greater the amount of oxygen transfer from slag
to metal for decarburisation. When the initial carbon content is below 200
ppm, there is no more oxygen pickup from the slag; the oxygen and
carbon contents decrease during degassing in accord with the stoichiometry of the reaction [C]+[O] → CO.

### 9.9.6 RATE OF DECARBURISATION

The rate of decarburisation is formulated as

$$\ln \frac{\%C_f}{\%C_i} = -k_C t \qquad (9.26)$$

where the rate constant $k_C$ for RH degassing is given by the following
relation.[19]

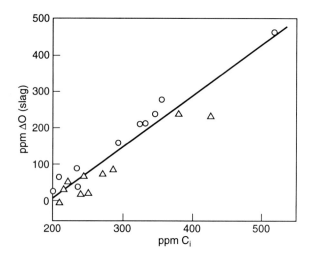

Fig. 9.35 Oxygen pickup from ladle slag during decarburisation in ($\Delta$) tank degasser and ($\bigcirc$) R H degasser.

$$k_C = \left(\frac{Q}{V}\right)\left(\frac{q}{Q+q}\right) \quad\quad\quad (9.27)$$

where $Q$ = circulation volume rate of molten steel, m³ min⁻¹

$V$ = volume of the steel in the ladle

$q$ = volumetric mass-transfer coefficient of decarburisation m³ min⁻¹.

For the 600 mm snorkel inside diameter and the argon injection rate of 2 Nm³min⁻¹ the rate constant $k_C$ = 0.12 min⁻¹. Kouroki *et al.*[20] and many other investigators have reported that the rate constant for RH decarburisation is not affected by the initial dissolved oxygen content of the steel. However, in most of the publications, no reference is made to the iron oxide content of the ladle slag. The rate constant $k_C$ appearing to be independent of the oxygen content of the steel may well be due to the use of a ladle slag that is high in iron oxide content. In fact, Ohma *et al.*[21] have found from the RH trial heats that the rate of decarburisation increases with an increasing iron oxide activity in the ladle slag. For example, $k_C$ is 0.1 min⁻¹ at $a_{FeO}$ = 0.1 and increases to 0.2 min⁻¹ at $a_{FeO}$ = 0.4. These findings support the view that with slag particles carried into the vacuum chamber, reaction (9.24) does participate in the vacuum decarburisation process.

The rate constant for decarburisation in the tank degasser[17,22] is about 0.07 min⁻¹ at the argon flow rate of 0.4 Nm³min⁻¹ and increases to about 0.09 min⁻¹ at 1.8 Nm³min⁻¹ for 180 to 200 t heats.

For the average rate constants $k_C$ = 0.12 min⁻¹ for RH degasser and 0.08 min⁻¹ for tank degasser, the extent of decarburisation in 20 minutes time of degassing would be as shown below.

| Initial %C | Final %C | |
| | RH degasser | Tank degasser |
| --- | --- | --- |
| 0.08 | 0.0072 | 0.016 |
| 0.06 | 0.0054 | 0.012 |
| 0.04 | 0.0036 | 0.008 |
| 0.02 | 0.0018 | 0.004 |

9.9.7 DEHYDROGENATION

To prevent the formation of hydrogen-induced hairline cracks or flaking in the bloom casting and thick slabs, the liquid steel is vacuum treated to lower the hydrogen content to 2 ppm or less. For the initial ($H_i$) and final ($H_f$) hydrogen contents, the rate equation for vacuum degassing is of the form

$$\ln \frac{[\text{ppm H}_f]}{[\text{ppm H}_i]} = -k_H t \tag{9.28}$$

The rate of dehydrogenation of 220 t heats at USS Gary–RH was evaluated by measuring the hydrogen content of the steel with the HYDRIS system during degassing.[23] The results obtained from the three heats (medium carbon, Al-killed steels) are shown in Fig. 9.36 for the 600 mm diameter snorkel and the steel circulation rate of about 140 t/min$^{-1}$ at USS Gary–RH; the rate constant $k = 0.13$ min$^{-1}$ obtained from these tests is in close agreement with the values given in publications on RH degassing

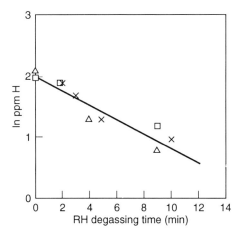

Fig. 9.36   Rate of dehydrogenation in RH degassing as measured by the HYDRIS system at U.S. Steel Gary Works. From Ref. 23.

from various steel plants. To achieve 2 ppm H in the steel, the RH degassing time is 7 min for $H_i$ = 5 ppm and 12.4 min for $H_i$ = 10 ppm.

Bannenberg et al.[13] derived a rate equation for dehydrogenation of 180 t steel in the tank degasser, as a function of tank gas pressure (1 to 100 mbar) and argon bubbling rate. The rate equation is for the mass transfer of hydrogen to rising argon bubbles in the melt as represented by equation (9.28). A close agreement was observed between the calculated and measured hydrogen contents after degassing times of 5 to 25 min; the HYDRIS system being used for the measurements. At 1 mbar tank pressure, $k_H$ = 0.09 min⁻¹ for the argon bubbling rate of 0.9 Nm³min⁻¹, increasing to 0.16 min⁻¹ at the flow rate of 1.8 Nm³min⁻¹; the latter is similar to $k_H$ in RH degassing with 1.5 to 2.0 Nm³min⁻¹ argon injection into the 600 mm diameter snorkel.

### 9.9.8 DENITROGENATION

Some nitrogen removal from liquid steel during vacuum degassing is possible, only if the steel is fully deoxidised and contains low concentrations of sulphur.

The rate equation derived by Bannenberg et al.[13] for denitrogenation of steel in the tank degasser is based on a mixed-control model; (i) liquid-phase mass transfer to argon bubbles and (ii) gas-metal interfacial reaction. The calculated data given in Fig. 9.37 for Al-killed steels, showing the effect of sulphur over the range 10 to 200 ppm S, were substantiated by the measured nitrogen contents.[13a]

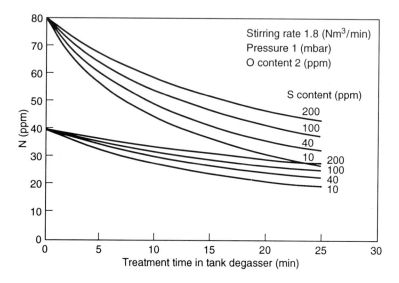

Fig. 9.37  Calculated sulphur effect on denitrogenation in tank degasser with argon stirring rate of 1.8 Nm³min⁻¹. From Ref. 13.

The rate of denitrogenation represented by equation (2.23) is re-written here in a simplified form

$$\frac{1}{\text{ppm N}} - \frac{1}{\text{ppm N}_o} = k_N (1 - \theta) t$$

where $k_N$ is the limiting apparent rate constant for O and S contents → zero. The calculated rate equation of Bannenberg *et al.* giving the results in Fig. 9.37 together with equation (2.21) for $(1 - \theta)$ give for the apparent rate constant $k_N = 0.0013$ (ppm N min)$^{-1}$ for 180 t heats at 1 mbar tank pressure and argon bubbling at the rate of 1.8 Nm$^3$min$^{-1}$. For much lower argon bubbling rates of 0.3 to 0.4 Nm$^3$min$^{-1}$, the value of $k_N$ will be much lower, e.g. probably in the range (4 to 5) × 10$^4$ (ppm N min)$^{-1}$.

## 9.10 CALCIUM TREATMENT

Most grades of steel are treated with calcium using either a Ca–Si alloy or a Ca–Fe(Ni) mixture, depending on the silicon specification. This treatment is made after trim additions and argon rinsing. In most melt shops, the cored-wire injection system is used in the calcium treatment of steel.

The melting and boiling points of calcium are 839°C and 1500°C respectively. The cored-wire containing Ca–Si or Ca–Fe(Ni) is injected into the melt at a certain rate such that the steel casing and contents of the cored-wire melt deep in the steel bath, e.g. 1.5 to 2.0 m below the melt surface without hitting the bottom of the ladle or resurfacing without being completely melted. In 200 to 240 tonne heats the cored-wire of 16 × 7 mm cross section or 13 mm diameter is injected at the rate of about 180 m min$^{-1}$.

During calcium treatment, the alumina and silica inclusions are converted to molten calcium aluminate and silicate which are globular in shape because of the surface tension effect. This change in inclusion composition and shape is known as the inclusion morphology control. This is demonstrated schematically in Fig. 9.38.

The calcium aluminate inclusions retained in liquid steel suppress the formation of MnS stringers during solidification of steel. This change in the composition and mode of the precipitation of sulphide inclusion during solidification of steel is known as sulphide morphology or sulphide shape control.

### 9.10.1 OBJECTIVES OF INCLUSION MORPHOLOGY CONTOL

Several metallurgical advantages are brought about with the modification of composition and morphology of oxide and sulphide inclusions by calcium treatment of steel, as for instance

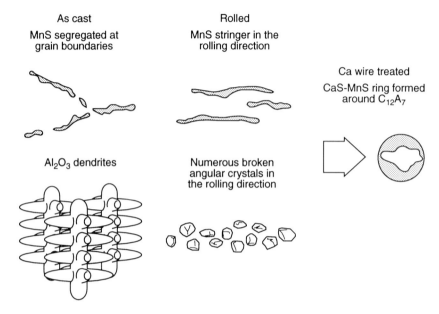

Fig. 9.38  Schematic illustration of modification of inclusion morphology with calcium treatment of steel.

(i)   to improve steel castability in continuous casting, i.e. minimise nozzle blockage,
(ii)  to minimise inclusion related surface defects in billet, bloom and slab castings,
(iii) to improve steel machinability at high cutting speeds and prolong the carbide tool life,
(iv)  to minimise the susceptibility of steel to re-heat cracking, as in the heat-affected zones (HAZ) of welds,
(v)   to prevent lamellar tearing in large restrained welded structures,
(vi)  to minimise the susceptibility of high-strength low-alloy (HSLA) linepipe steels to hydrogen – induced cracking (HIC) in sour gas or sour oil environments.
(vii) to increase both axisymmetric tensile ductility and impact energy in the transverse and through-thickness directions in steels with tensile strengths below 1400 MPa.

9.10.2 REACTION OF CALCIUM IN LIQUID STEEL

When calcium is injected deep into the melt as an alloy of Ca–Si, Ca–Al or as pure Ca admixed with nickel or iron powder, the subsequent reactions in liquid steel will be the same. The following series of reactions are expected to occur to varying extents in Al-killed steels containing alumina inclusions.

$$Ca\,(l) \rightarrow Ca(g)$$
$$Ca(g) \rightarrow [Ca]$$
$$[Ca] + [O] \rightarrow CaO$$
$$[Ca] + [S] \rightarrow CaS$$
$$[Ca] + (x+\tfrac{1}{3})Al_2O_3 \rightarrow CaO{\cdot}xAl_2O_3 + \tfrac{2}{3}[Al] \qquad (9.29)$$

Depending on the steel composition, the manner of calcium injection and other process variables, there will be variations in the conversion of alumina inclusions to aluminate inclusions; the smaller inclusions will be converted to molten calcium aluminate more readily than the larger inclusions.

As is seen from the plant data in Fig. 9.39, the deoxidation of steel with aluminium is shifted to lower levels of residual dissolved oxygen upon subsequent calcium treatment because of the lower alumina activity in the modified inclusions. It is a generally accepted view that the calcium retained in liquid steel is primarily in the form of calcium aluminate inclusions, and to a lesser extent as calcium sulphide inclusions dispersed in the melt, or as dissolved Ca in the steel. The analytical data from various ladle refining shops are summarised in Fig. 9.40 for Al-killed steels containing S < 50 ppm and S > 100 ppm. In all cases, the calcium treatment was made by the cored-wire injection of Cal–Sil at the rates of 0.16 to 0.36 kg Ca/tonne steel, with the 16 × 7 mm cross section cored-wire being injected at the rate of about 180 m/min.

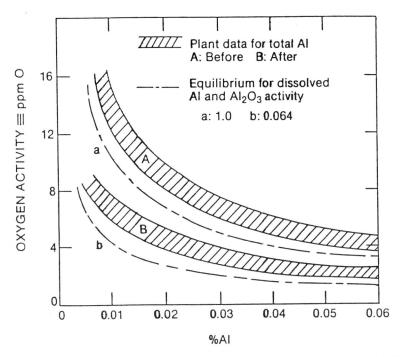

Fig. 9.39   Oxygen sensor readings before and after Cal–Sil injection into Al-killed steel at 1600 ± 15°C. From Ref. 24.

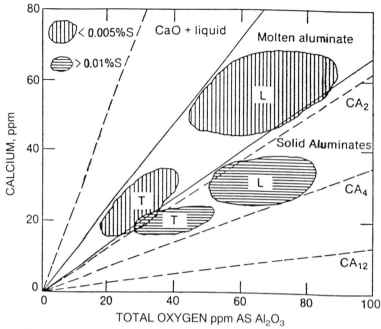

Fig. 9.40   Relation between calcium and total oxygen contents of Al-killed steels after Cal–Sil treatment. From Ref. 24.

The dotted lines for the indicated compositions of calcium aluminates ($CA_x$) are calculated from the stoichiometry of the reaction (9.29) and the phase equilibrium diagram for the system $CaO$–$Al_2O_3$; the region for molten calcium aluminate is for 1600°C. The shaded areas marked L are for the steel samples taken a few minutes after Cal–Sil injection, those marked T are for the tundish samples. Because of the floatation of inclusions out of the melt, the total calcium and oxygen contents of the steel in the tundish samples are always lower than those in the ladle soon after the end of Cal–Sil injection. For a given total oxygen content of the steel, initially as alumina inclusions, a greater amount of calcium is retained in the low-S steels. In liquid steels containing S < 50 ppm, there is almost complete conversion of alumina inclusions to molten calcium aluminate inclusions. In high-sulphur steels the conversion is partial, resulting in solid aluminate, presumably coated with a thin layer of molten aluminate, or solid aluminate with unreacted alumina core. Some calcium sulphide is also associated with these aluminate inclusions.

9.10.3 EFFICIENCY OF CALCIUM USAGE AND RETENTION IN STEEL

The material balance for calcium consumption is given by the following summation:

amount injected $W = W_{sol} + W_{OS} + W'_{OS} + W_{sl} + W_v$

where  $W_{sol}$  = amount dissolved in liquid steel
$W_{OS}$  = amount retained as aluminate and sulphide inclusions
$W'_{OS}$  = amount reacted with sulphur and alumina and floated out
$W_{sl}$   = amount reacted with the ladle slag
$W_v$    = amount burnt at the surface of the melt

The study of plant data on calcium treated steels and the author's theoretical considerations indicate that

$$W_{sol} \ll W_{OS}$$

With this simplification, the efficiency of calcium usage is approximated by

$$E_u = \frac{W_{OS} + W'_{OS}}{W} \tag{9.30}$$

The efficiency factor for calcium retention in liquid steel would be

$$E_r = \frac{W_{OS}}{W} \tag{9.31}$$

Experience has shown that the efficiency factor for calcium retention decreases with an increasing amount of calcium injection beyond a certain amount, depending on the total oxygen content of the steel as oxide inclusions. Because of inclusion floatation out of the melt during the time interval prior to casting, $(E_r)_T$ in the tundish will be lower than $(E_r)_L$ in the ladle soon after calcium treatment. As a rule of thumb, we may take $(E_r)_T \approx \frac{1}{2}(E_r)_L$.

From many trial heats the following values of calcium retention are obtained for Al-killed steels initially containing 50 to 80 ppm O) as alumina inclusions.

| kg Ca/tonne injected | $(E_r)_L$ | $(E_r)_T$ |
|---|---|---|
| 0.16 | 0.24 to 0.30 | 0.12 to 0.15 |
| 0.36 | 0.12 to 0.18 | 0.06 to 0.09 |

If the steel is to be degassed for hydrogen removal, the calcium treatment should be made after vacuum degassing.

## 9.10.4 SULPHIDE MORPHOLOGY (SHAPE) CONTROL

In steels not treated with calcium, the sulphur precipitates as minute particles of MnS in the last interdendritic liquid to freeze, which delineate the prior austenite grain boundaries in the as cast structure. During hot rolling, the MnS particles are deformed forming stringers in the rolled product. The MnS and $Al_2O_3$ stringers make steel susceptible to, for example (i) re-heat cracking in the heat-affected zones of welds, (ii) lamellar tearing in large restrained welded structures and (iii) hydrogen-induced cracking in sour gas or oil environments.

   In the calcium treated low-sulphur steels, the grain boundary precipitation of MnS is suppressed by the precipitation of sulphur as a Ca(Mn)S complex on the calcium aluminate inclusions during solidification of steel, by the following reaction.

$$(CaO) + 2[S] + [Mn] + \tfrac{2}{3}[Al] \rightarrow (CaS \cdot MnS) + \tfrac{1}{3}(Al_2O_3) \qquad (9.32)$$

   The extent of sulphide-shape control that can be achieved during solidification of the calcium treated steel is governed by the total oxygen, sulphur and calcium contents of the steel. From the theoretical consideration of reactions occurring in the impurity enriched interdendritic liquid during solidification, the following criteria are derived for the tundish compositions of the Al-killed steels, to achieve an acceptable degree of sulphide-shape control during solidification.

As aluminate
inclusions

| ppm O | ppm Ca | % Mn | ppm S |
|-------|--------|---------|-------|
| 25 | 20–30 | 0.4–0.6 | < 20 |
| 25 | 20–30 | 1.3–1.5 | < 30 |
| 12 | 15–20 | 0.4–0.6 | < 10 |
| 12 | 15–20 | 1.3–1.5 | < 15 |

   If an acceptable level of sulphide shape control is not achieved, the HSLA steels for linepipes and off shore installations will be susceptible to hydrogen-induced cracking (HIC) in sour gas (containing $H_2S$) and sour oil environments. The general concensus that emerged from many experimental studies is that a high resistance to HIC in HSLA steels can be assured, by lowering the sulphur and oxygen ($Al_2O_3$ inclusions) contents of the steel and by treating steel with calcium to prevent the formation of MnS during solidification.

## 9.11 INCLUSION CONTROL IN BEARING STEELS

The deleterious effects of oxide inclusions on the fatigue resistance of bearing steels are now well understood through many years of experimental and theoretical studies. This subject was well documented in a series of papers by Brooksbank and Andrews.[25] The fatigue failure is a consequence of the structural tessellated stresses in and around the inclusions.

As originally conceived by Laszlo,[26] the tessellated stresses are of the general form

$$\text{Stress} = \pm \, \Phi \, [(\alpha_m - \alpha_i)\Delta T] \qquad (9.33)$$

where $\Phi$ is a variable function depending on (i) the elastic moduli of the inclusions and the steel matrix, (ii) the inclusion size, shape, and distribution, and (iii) the position and direction of the individual stress considered. The sign depends on the type of stress, e.g., positive for circumferential and negative for radial. Laszlo termed $(\alpha_m - \alpha_i)\Delta T$ the 'strain potential of structural tessellation,' where $\alpha_m$ is the mean linear coefficient of thermal expansion of the matrix, $\alpha_i$ is the mean linear coefficient of the inclusion, and $\Delta T$ is the temperature change in the system.

Upon cooling, if the contraction of inclusion is less than that of the steel matrix, i.e., $\alpha_i < \alpha_m$, stresses will develop in and around the inclusions leading to crack initiation, hence the deterioration of the fatigue life. Inclusions having $\alpha_i > \alpha_m$ are not detrimental to fatigue properties because adverse stresses will not develop. Brooksbank and Andrews have shown that the variations in fatigue life with the type of inclusion in the steel are in general accord with those predicted from the term $(\alpha_m - \alpha_i)\Delta T$, as shown in Fig. 9.41 for oil-quenched 1 percent C–Cr steels. The most detrimental inclusions to fatigue life are: calcium aluminates, alumina, spinels, silicates and nitrides in this descending order. In the case of sulphide inclusions, the coefficient of contraction $\alpha_i$ is greater than that of the steel, consequently they are non-detrimental to fatigue life; MnS inclusions are the most innocuous to fatigue life. It is also known that encapsulating alumina inclusions with a layer of MnS, occurring during solidification of the steel, reduces the detrimental effect of alumina. According to the theoretical analysis of Brooksbank and Andrews, the mass ratio O/S in the steel should not exceed 0.4 to eliminate stresses around $Al_2O_3$ by the encapsulating MnS. This theoretical prediction has been substantiated by many fatigue tests on bearing steels, as cited in references (27) and (28). To attain adequate encapsulation of alumina inclusions by MnS, the sulphur content of many bearing steels is maintained on the high side, e.g. 0.015 to 0.025 percent S.

Although the encapsulation of calcium aluminate particles by Ca(Mn)S will soften their detrimental effect on fatigue life, it is not as effective as encapsulation of $Al_2O_3$ particles by MnS. Therefore, care is taken in the ladle refining of bearing steels to prevent the entrainment of calcium-bearing ladle slag in the steel.

In the past, the vacuum-carbon deoxidation (VCD) was mandatory for the ladle refining of bearing steels. However, with the much improved ladle refining facilities we have today and with careful implementation of the 'clean-steel practice' in aluminium deoxidation of steel (without the initial VCD), it now is possible to lower the concentration of $Al_2O_3$ inclusions in the steel below 15 ppm. At this low level of $Al_2O_3$ content,

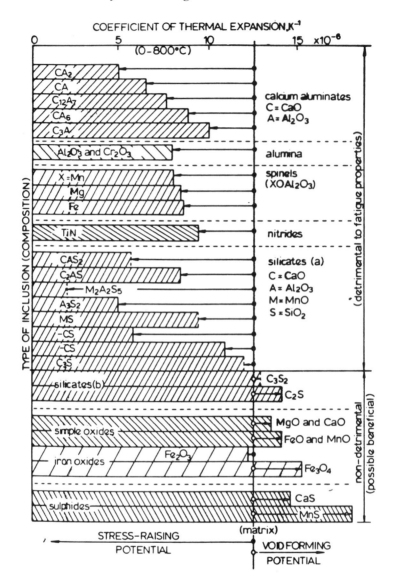

Fig. 9.41    Stress-raising and void forming properties of inclusions in 1% C–Cr bearing steels. From Ref. 25.

coupled with encapsulation by MnS, the bearing steel will have the desired resistance to fatigue failure.

### 9.12 OTHER REFERENCES ON INCLUSIONS IN STEEL

It is outside the scope of this book to make a comprehensive presentation on the subject of inclusions in steel. For indepth background information on all metallurgical aspects of inclusions in steel, reference should be made to a series of books authored by Kiessling under the heading *Non-metallic inclusions in steel* Parts I–IV over a period 1964–1976, which were subsequently compiled as a single book[29] by The Metals Society; Part V of this series by Kiessling was published in 1989. References may also be made to a series of papers published in various conference proceedings on the subject of inclusions in steel[30–32].

The physicochemical aspects of inclusions in steel have been studied in detail experimentally and theoretically by the research personnel in IRSID since the early 1980s. Their assessment of morphology modification of inclusions by the calcium treatment of steel is based on their experimentally determined solubility products of CaO and CaS in liquid steel.[33] As discussed in section 4.4 of Chapter 4, these measured solubility products are several orders of magnitude greater than those derived from the thermochemical data. However, not being deterred by this dilemma, due acknowledged consideration is given here to the consensus of Gaye *et al.*[34] on the state of CaS–calcium aluminate equilibrium in liquid steel as represented by the equilibrium diagram for 1600°C in Fig. 9.42. For dilute solution in liquid steel, the activities of dissolved Ca, S and Al are equivalent to their mass concentrations in percent as given in this diagram.

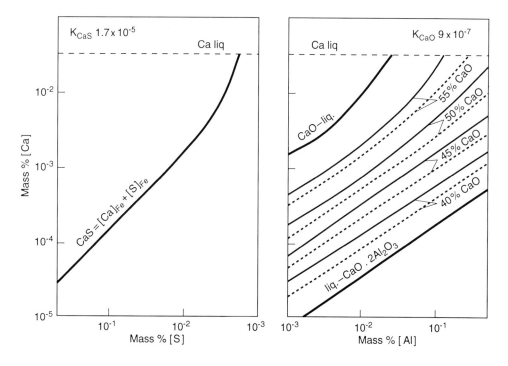

Fig. 9.42  Computed equilibrium diagram for Fe–Al–Ca–O–S system at 1600°C; (– – –) curves for aluminate inclusions containing 5% $SiO_2$. From Ref. 34.

On the basis of these equilibrium data, Gaye *et al.*[34] developed a computer programme to evaluate, from a global analysis of a steel sample, the composition and amounts of inclusions which are formed in liquid steel at the treatment temperature and during the subsequent cooling until the onset of solidification. The computed data in Fig. 9.43 show the precipitation path of inclusions formed during the calcium treatment and subsequent cooling of the liquid steel containing 0.02% S. The left hand diagram in Fig 9.43 represents the distribution among liquid steel, oxide and sulphide inclusions of the remaining total calcium analysed in the cast product. In the case considered with the steel containing 0.02% S, the CaS precipitates only during cooling. The right hand diagram depicts the CaO and $Al_2O_3$ contents of the inclusions formed during calcium treatment and cooling of the steel (solid-line curves) as well as the average contents in the inclusion population (dotted curves). The computed average composition of these oxide inclusions is 51.6% CaO, 38.8% $Al_2O_3$, 9.5% $SiO_2$ and 0.1% MnO. Additional computed and analytical data are presented in another paper by Gaye *et al.*[35] on the control of inclusion composition by slag treatment of steel and inclusion transformations during cooling of the solidified steel.

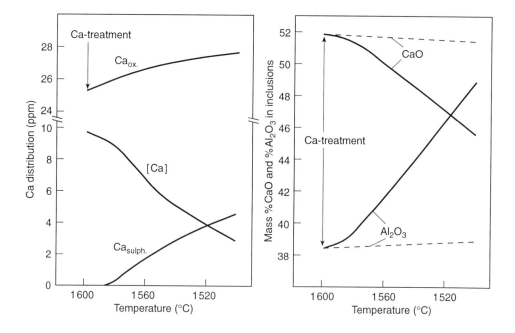

Fig. 9.43   Computed precipitation path of inclusions in Ca-treated medium carbon steel: 0.38% C, 0.75% Mn, 0.25% Si, 0.015% Al, 35 ppm Ca and 30 ppm O. From Ref. 34.

## REFERENCES

1.  D.M. SCHRADER and L.L. RANKOVIC, *Iron and Steelmaker*, 1988, **15**(6), 23.
2.  C.E. TOMAZIN, E.A. UPTON and R.A. WALLIS, *Steelmaking Conf. Proc.*, 1986, **69**, 223.
3.  *The Making, Shaping and Treating of Steel*, U.S. Steel, Published by Association of Iron and Steel Engineers 1985.
4.  B. GRABNER and H. HOEFFGEN, *Radex Rundschau*, 1985, **3**, 581.
5.  P.E. ANAGBO and J.K. BRIMACOMBE, *Iron & Steelmaker*, 1988, **15**(10), 38; **15**(11), 41.
6.  W. PLUSCHKELL, *Stahl und Eisen*, 1981, **101**, 97.
7.  D. MAZUMDAR and R.I.L. GUTHRIE, *Metall. Transaction B*, 1986, **17B**, 725.
8.  J. ISHIDA, K. YAMAGUCHI, S. SUGIURA, K. YAMANO, S. HAYAKAWA and N. DEMUKAI, *Denki-Seiko*, 1981, **52**, 2.
8a. A.W. HUTNIK and J.B. HEMPHILL, *National Open Hearth & Basic Oxygen Conf. Proc.* 1974, **57**, 358.
8b. R.J. HAWKINS, *ibid*, 1976, **59**, 479.
8c. J.H. MIKULECKY, *ibid*, 1977, **60**, 338.
9.  T. CHOH, K. IWATA and M. INOUYE, *Trans. Iron and Steel Inst. Japan*, 1983, **23**, 598.
10. E.T. TURKDOGAN, *Ironmaking and Steelmaking*, 1988, **15**, 311.
11. M.R. RILEY and L.G. NUSSELT, *Proc. 5th Inter. Iron and Steel Cong.*, 1986, **6**, 177.
12. A. THOMAS, F. VILLETTE and J. PITON, *Iron and Steel Engineer*, 1986, Feb., 45.
13. N. BANNENBERG, B. BERGMANN and H. GAYE, *Steel Research*, 1992, **63**(10), 431.
13a. N. BANNENBERG and B. BERGMANN, *Stahl u. Eisen*, 1992, **112**(2), 57.
14. B. OZTURK and E.T. TURKDOGAN, *Metal Science*, 1984, **18**, 306.
15. S. GILBERT, G.G. MONOS and E.T. TURKDOGAN, *Steelmaking Conference Proceedings*, 1988, **71**, 291.
16. Y. OGUCHI, T. FUJII, H. KAMAMURA, F. SUDO, A. NANBA and M. ONISHI, *Tetsu-to-Hagané*, 1983, **69**, A.37.
17. D. NOLLE, U. EULENBURG, A. JAHNS and H. MISKA, *Inter. Conf. Secondary Metallurgy*, ICS, 1987, p. 269.
18. U.S. Steel Technical Center, private communication.
19. N. SUMITA *et al.*, Kawasaki Steel Gihou, 1983, **15**, 152.
20. S. KOUROKI, T. OBANA, Y. SHIROTA and M. TANAKA, Scaninject V, 1989, p. 642.
21. H. OHMA, H. NAKATA, K. MORI and T. YAJIMA, *Steelmaking Conf. Proc.*, 1986, **69**, 327.
22. N. BANNENBERG, PH. CHAPELLIER and M. NADIF, *Stahl u. Eisen*, 1993, **113**(9), 75.
23. R.P. STONE, J. PLESSERS and E.T. TURKDOGAN, *Stahl u. Eisen*, 1990, **110**(11), 65: *Electric Furnace Conf. Proc.*, 1990, **48**, 83.
24. E.T. TURKDOGAN, in *Calcium Treatment Symposium*, p. 3. The Institute of Metals (now The Institute of Materials), London, 1988.
25. D. Brooksbank and K.W. ANDREWS, *J. Iron Steel Inst.*, 1969, **207**, 474; 1970, **208**, 582; 1972, **210**, 246 and 765.
26. F. LASZLO, *J. Iron Steel Inst.*, 1943, **147**, 173; 1945, **152**, 207.

27. R. Tricot, in *Production and Application of Clean Steels*, Pub. No. 134, The Iron and Steel Institute, London, 1972, 199.
28. E. Emekes, in *Production and Application of Clean Steels*, Pub. No. 134, The Iron and Steel Institute, London, 1972, 215.
29. R. Kiessling, *Non-metallic inclusions in steel*, Parts I–IV, The Metals Society, London, 1978; Part V, The Institute of Metals (now The Institute of Materials), London, 1989.
30. *Inclusions*, F.B. Pickering, ed., Institution of Metallurgists, London (now The Institute of Materials), Monograph no. 3, 1979.
31. *Swedish symposium on non-metallic inclusions in steel*, H. Nordberg and R. Sandstrom, eds., Swedish Institute of Metal Research, Stockholm, 1981.
32. *Clean Steel 4 Proc.*, The Institute of Materials, London, 1994.
33. M. Nadif and C. Gatellier, *Rev. Metall. CIT*, 1986, **83**, 377.
34. H. Gaye, C. Gatellier, M. Nadif, P.V. Riboud, J. Saleil and M. Faral, in *Clean Steel 3*, p.137, The Institute of Metals (now The Institute of Materials), London, 1986.
35. H. Gaye, C. Gatellier and P.V. Riboud, in Proc. *The Ethem T. Turkdogan Symposium*, p.113. The Iron and Steel Society, Inc., Warrendale, PA, USA, 1994.

# CHAPTER 10

# Reactions during Steel Solidification

Problems sometimes encountered in the steel castability, hot workability, machinability and shortcomings of the mechanical properties of the product, are caused by common inpurity elements in the steel such as oxygen, sulphur, nitrogen and hydrogen. These impurities manifest their effects in a variety of ways during solidification of the steel, hot working and machining of the product and in service performance of the product. The effects of these impurities on various properties of the steel are accentuated by the solidification process involving interdendritic microsegregation. Some special cases discussed in this chapter are typical examples of the consequences of microsegregation occurring during solidification of the steel; namely (i) the formation of subsurface blowholes, (ii) the mode of formation of oxysulphide inclusions in free-machining steels, (iii) the critical ratio $(Mn/S)_c$ to suppress hot shortness of steel, (iv) the sulphide shape modification and (v) the control of nitride precipitation in continuous casting.

## 10.1 SOLUTE ENRICHMENT IN INTERDENDRITIC LIQUID STEEL

For background information on the technical aspects of the morphology of dendritic solidification of metal alloys, reference may be made to the work of Chalmers,[1-3] Flemings[4,5] and their co-workers. The dendrites and sub-dendritic branches have a platelike morphology; as they grow in the direction of heat flow, subdendritic branches are formed such that liquid becomes trapped between the interdendritic branches of the solidifying metal.

### 10.1.1 SCHEIL EQUATION

Microsegregation resulting from solute rejection during dendritic freezing in binary alloys was formulated by Scheil[6] for two limiting cases summarised below.

A. Complete solute diffusion in the solid (dendrite arms) and interdendritic liquid, appropriate for the interstitial solutes C, O, N and H.

$$C_l = \frac{C_o}{k + (1 - k)g} \qquad (10.1)$$

297

where  $g$  = volume fraction remaining liquid, in any given volume element of the solidifying mass
$C_o$ = initial solute concentration at $g = 1$
$C_l$ = solute concentration in the enriched liquid
$k$  = solid/liquid equilibrium solute distribution ratio.

B. Negligible diffusion in the solid and complete diffusion in the interdendritic liquid, appropriate for substitutional elements, Mn, Si, Nb, P . . .

$$C_l = C_o g^{k-1} \tag{10.2}$$

The solute distribution ratios in dilute solutions of $X$ in selected systems Fe–$X$ are listed below.

| Solute $X$ | $k = [\%X]_\delta/[\%X]_l$ |
|:---:|:---:|
| Al | 0.60 |
| C | 0.20 |
| H | 0.27 |
| Mn | 0.74 |
| N | 0.27 |
| Nb | 0.27 |
| O | ~ 0 |
| P | 0.08 |
| Si | 0.60 |
| S | 0.04 |
| Ti | 0.60 |

An example is shown in Fig. 10.1 of solute enrichment in the interdendritic liquid for steel containing Mn, Si, C and O. It should be remembered that the term 'percent local solidification' refers to the progress of solidification in any given volume element of the solidifying casting.

### 10.1.2 REACTIONS IN THE INTERDENDRITIC LIQUID

When solute concentrations in the interdendritic liquid steel become sufficiently high, reactions may occur between the solutes resulting in the formation of oxide, silicate, oxysulphide inclusions, and in some cases, gas bubbles. When interdendritic reactions occur, the solute enrichment is computed by incorporating the Scheil equation with equations for reaction equilibria, on the assumption that no supersaturation is needed for the formation of the reaction products. Methods of computations are given in the author's previous publications.[6–8]

An example of the calculations is given in Fig. 10.2 showing the concentration of dissolved oxygen in the interdendritic liquid steel controlled by the local deoxidation reaction for steel containing 0.50% Mn, 0.004% O and zero to 0.030% Si. If there were no deoxidation reaction occurring at

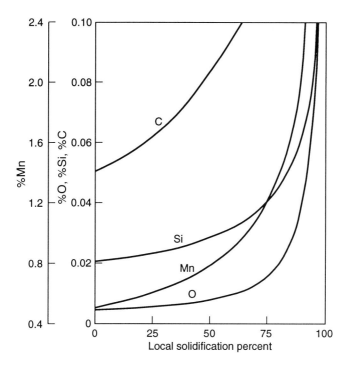

Fig. 10.1   Interdendritic solute enrichment, if no reaction occurs between them during solidification of the steel.

90% local solidification, the concentration of dissolved oxygen would have been enriched to 0.04% O. In steel containing as little as 0.01% Si, the Si/Mn deoxidation in the interdendritic liquid suppresses oxygen enrichment. The equilibrium oxygen contents of the interdendritic liquid steel are shown in Fig. 10.3 for the C–O reaction at 1 atm CO at various stages of solidification for several initial carbon concentrations. From the comparison of the oxygen concentration curves in Figs. 10.2 and 10.3, we see that for steel containing 0.5% Mn, 0.03% Si and 0.004% O (dissolved), the C–O reaction generating CO bubbles will not occur at any stage of the solidification in steels containing less than 0.08% C and no H or N.

Another important reaction in the impurity enriched interdendritic liquid is the precipitation of MnS inclusions. At the liquidus temperatures of 1525 to 1500°C, the solid MnS solubility product [%Mn][%S] is 1.07 to 0.92 respectively. In steel with 1.0% Mn, the product [%Mn][%S] in the interdendritic liquid increases during solidification as shown in Fig. 10.4 for steels containing 0.01 and 0.03% S. For an average liquidus temperature of 1510°C and under equilibrium conditions, the MnS precipitation begins when the concentration product [%Mn][%S] reaches the equilibrium solubility product of about 1.0.

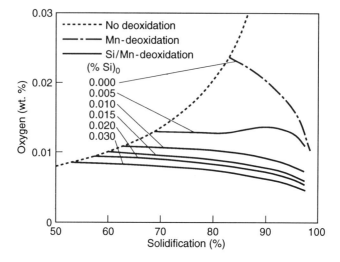

Fig. 10.2   Oxygen content of enriched liquid controlled by Si–Mn deoxidation during solidification. From Ref. 6.

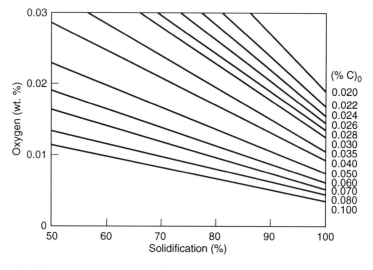

Fig. 10.3   Equilibrium oxygen content of steel, for carbon–oxygen reaction at 1 atm. CO at various stages of solidification for several initial carbon concentrations. From Ref. 6.

The interdendritic precipitation of MnS in the later stages of local solidification is in evidence, as shown in Fig. 10.5, from the photomicrogaph of the etched section of an Al-killed steel casting containing 1.5% Mn, 0.25% C and 0.05% S. In this as-cast structure, the circular or oblong shaped white regions are the cross sections of the dendrite arms; the darker regions correspond to the higher manganese contents, hence the pearlitic structure upon the austenite to ferrite phase transformation on cooling. The MnS inclusions are lined up along the middle of the interdendritic region.

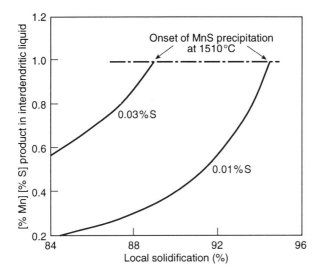

Fig. 10.4 Estimation of the onset of MnS precipitation in the impurity enriched interdendritic liquid steel initially containing 1.0% Mn and 0.01 or 0.03% S.

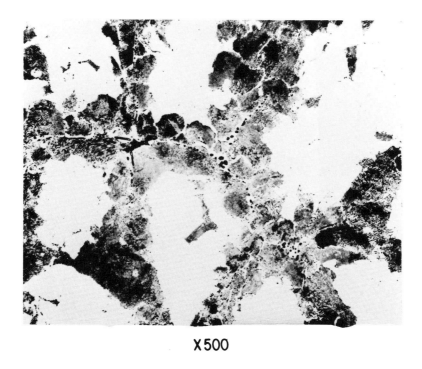

Fig. 10.5 Manganese sulphide inclusions in the interdentritic regions in the as-cast conditions (0.25% C, 1.5% Mn and 0.05% S). From Ref. 7.

### 10.1.3 LIQUIDUS TEMPERATURE AND SOLUTE ACTIVITY COEFFICIENTS IN THE INTERDENDRITIC LIQUID

As the interdendritic liquid steel becomes enriched with alloying elements, the liquidus temperature decreases with the progress of local solidification. Because the equilibrium constants of reactions are temperature dependent, in calculating the state of reactions in the enriched interdendritic liquid, due account should be taken of the liquidus temperature of the steel using the formulation given in section 4.3.2b. Also due account should be taken of the activity coefficients of solutes in the alloy enriched interdendritic liquid using the interaction coefficients given in Table 4.2.

As an example, let us consider a steel containing 1.0% Mn, 0.2% Si. 0.5% C, 0.02% S and 0.01% P. At 90% local solidification, i.e. $g = 0.1$, the enriched liquid will contain: 1.82% Mn, 0.63% Si, 1.78% C, 0.182% S and 0.083% P. For the initial steel composition, the liquidus temperature is about 1492°C; in the enriched liquid at $g = 0.1$, the liquidus temperature is 1404°C. Changes in the solute activity coefficients from the initial ($g = 1$) to $g = 0.1$ are as follows.

| Solidification stage, $g$ | $f_{Mn}$ | $f_{Si}$ | $f_C$ | $f_S$ | $f_N$ | $f_H$ |
|---|---|---|---|---|---|---|
| 1.0 | 0.92 | 1.29 | 1.19 | 1.11 | 1.15 | 1.07 |
| 0.1 | 0.74 | 2.56 | 1.94 | 1.58 | 1.69 | 1.34 |

## 10.2 SUBSURFACE BLOWHOLE FORMATION

In continuous casting the subsurface blowholes may form if the concentrations of C, O, N and H are high. The photograph of a bloom section in Fig. 10.6 shows a severe case of subsurface blowholes that occurred in continuous casting. During reheating to the hot-rolling temperatures, the oxidising furnace gases penetrate below the surface through the blowholes. The subsurface scale thus formed cannot be removed by the scale breaker; consequently during hot rolling, defects are developed on the surface of the slab. In the case of rephosphorised and resulphurised steel, the formation of subsurface scale is accelerated by the oxidation of phosphorus in the steel to iron phosphate, or by the formation of liquid iron-manganese oxysulphide inclusions, thus increasing the extent of surface defects on the hot-rolled product.

In the early 1960s, the author developed a simple theoretical model to estimate steel compositions that will insure a non-porous surface in the continuous casting of billets, blooms or slabs[6]. The practical validity of this simple theoretical model was subsequently substantiated experimentally for wide ranges of steel compositions.[7,9-11].

Fig. 10.6 Section of a continuous cast 260 mm bloom showing subsurface blowholes.

The solutes dissolved in liquid steel that contribute to the formation of blowholes are H, N and CO. When the sum of the equilibrium partial pressures of these solutes in the enriched interdendritic liquid exceeds the local external pressure, gas bubbles will be generated, resulting in expulsion of the interdendritic liquid into the neighbouring regions, hence the formation of blowholes or pinholes. This statement is summarised by the following equation depicting the onset of subsurface blowholes.

$$p_{H_2} + p_{N_2} + p_{CO} > P_s + P_f + \frac{2\sigma}{r} \qquad (10.3)$$

where  $P_s$ = atmospheric pressure on the surface of liquid steel in the mould

$P_f$ = ferrostatic pressure at the location of the blowholes

$\sigma$ = surface tension of liquid steel in contact with the gas bubble of radius $r$.

For a gas bubble of 1-mm radius, the excess pressure due to the surface tension effect is only about 0.02 to 0.03 atm. For the size of pinholes or blowholes observed in castings, the surface tension effect becomes insignificant. The subsurface blowhole formation in continuous casting will occur in the early stages of the solidification, a short distance below the meniscus where the total pressure is only slightly above atmospheric. We may take an average of 1.05 atm as the critical total gas pressure for the onset of the subsurface blowhole formation in continuous casting. The gas partial pressures are calculated for the solute enriched concentrations using the solubility data given in Chapter 4. The total gas pressure buildup at 90 percent local solidification is used in estimating steel compositions which are susceptible to the formation of blowholes.

Two examples are given in Fig. 10.7 showing the effective gas partial pressure buildup in the interdendritic liquid at about 90% of solidification. Fig. 10.7a is for Al-killed steels containing C < 0.1% and 0, 50 and 100 ppm N. Hydrogen in the steel is a major contributor to the formation of subsurface blowholes. With the Al-killed steel containing 50 ppm N, the critical hydrogen content is 8 ppm for the onset of blowholes. In fact, if the steel in the tundish contains about 9 ppm H, which sometimes can be the case in Q–BOP steelmaking, a breakout in the caster becomes imminent. Gas bubbles forming in the early stages of solidification enter the mould gap and become intermixed with mould flux, thus lowering the thermal conductivity at the gap; this situation can be the cause of a break-out in the

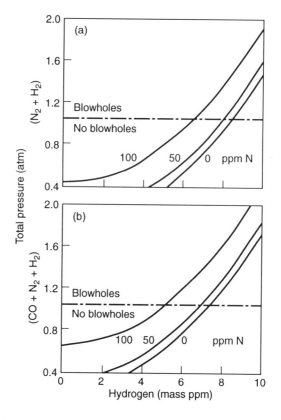

Fig. 10.7   Total gas pressure at 90% local solidification. (a) Al-killed steel containing < 0.1% C (b) Semi-skilled steel containing: (i) 1.5% Mn, 0.2% Si, 0.08% C, 10 ppm O (ii) 0.5% Mn, 0.03% Si, 0.02% C, 40 ppm O.

mould. Fig. 10.7b is for semi-killed steels containing (i) 1.5% Mn, 0.2% Si, 0.08% C and 10 ppm O and (ii) 0.5% Mn, 0.03% Si, 0.02% C and 40 ppm O. The critical hydrogen content for the onset of blowhole formation is 4 ppm H for 100 ppm N and 6 ppm H for 50 ppm N in the steel.

## 10.3 CASTABILITY OF RESULPHURISED STEEL

Steel compositions best suited for free machinability must also satisfy the requirements for castings free of blowholes. Unfortunately, the metallurgical requirements for good machinability are often contrary to other metallurgical requirements for good castability. For instance, steel compositions for a high machinability rating, e.g. AISI Grade 1215 – resulphurised and Grade 12L14 – resulphurised and leaded steel, developed originally for ingot mould casting, are not suitable for continuous casting because of the low silicon (<0.01% Si) and high oxygen contents (250 to 350 ppm O) of the steel.

In the resulphurised steel (~ 0.3% S), to which no Si or Al has been added, the primary reaction in the interdendritic liquid is the formation of liquid manganese oxysulphide particles.

$$[Mn] + [O] \rightarrow (MnO)$$
$$[Mn] + [S] \rightarrow (MnS) \quad \text{liquid oxysulphide} \quad (10.4)$$

From the known thermodynamic data on the MnO–MnS system[8] and the solute enrichment formulations, the formation of oxysulphides during steel solidification are computed for an average liquidus temperature of 1500°C. Calculations are made for resulphurised steel (0.30% S, 200 ppm

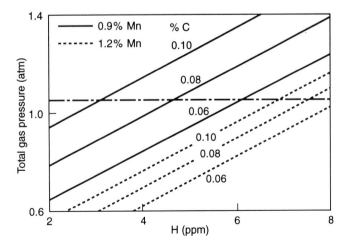

Fig. 10.8  Computed gas pressure in interdendritic liquid steel at 90% local solidification for steels containing 0.3% S, 50 ppm N and indicated concentrations of Mn, C and H.

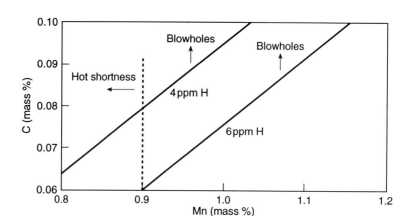

Fig. 10.9   Estimated C and Mn contents of free-machining steels (containing 0.3% S, 200 ppm O, 50 ppm N and 4 or 6 ppm H) for continuous casting, free of subsurface blowholes.

O) containing 0.9 to 1.2% Mn, 0.06 to 0.10% C, 50 ppm N and 3 to 8 ppm H. The computed total gas pressure in the interdendritic liquid at 90% local solidification is shown in Fig. 10.8. Taking 1.05 atm as the critical pressure for the onset of blowhole formation, the critical carbon and manganese contents are estimated as given in Fig. 10.9. With carbon contents above the line, the subsurface blowholes are expected to form in continuous casting.

It should be noted that the fast rate of cooling inherent to continuous casting, results in the formation of oxysulphide particles that are smaller in size and have Mn/Fe ratios lower than those formed in the ingot mould casting. Consequently, for the same steel composition the machinability rating of continuously cast steel will be lower than that which is cast in a large ingot mould. This is demonstrated by the machinability ratings given below.

| Cross section of casting, mm | Average size of MnS inclusions, μm$^2$ | Screw test Machinability, parts per hour* |
|---|---|---|
| 152×152 billet | 39 | 90 |
| 254×254 bloom | 62 | 145 |
| 380×508 bloom | 80 | 170 |
| 810×810 ingot | 150 | 300 |

*Tests were made at the U.S. Steel Technical Center

## 10.4 CRITICAL RATIO $(Mn/S)_c$ TO SUPPRESS HOT SHORTNESS OF STEEL

The hot shortness of steel at hot working temperatures and crack susceptibility of steel in continuous casting, have long been known to be due, in part, to the presence of liquid oxysulphide and iron sulphide at the grain boundaries. The unanimous consensus of many plant investigations and laboratory studies, is that the hot ductility decreases with a decreasing ratio of Mn/S below some critical ratio. There are wide variations in the technical literature on the value of the critical ratio $(Mn/S)_c$ for the suppression of hot shortness of steel. These reported variations are due to the method of testing and to the criterion chosen for the definition of the ductile/brittle transition.

The results of three separate studies[8,12,13] are given in Fig. 10.10 for Al-killed steels, showing the relationship between the critical ratio $(Mn/S)_c$ and sulphur content of the steel. For compositions above the line only MnS forms; below the line some liquid FeS forms with the FeS/MnS ratio increasing with a decrease in the Mn/S ratio below the critical value. The points in Fig. 10.10 are based on the experimental work of Schmidtmann and Rakoski[14] as interpreted ty Toledo *et al.*[13] The formulation derived by the author[8] for the estimation of $(Mn/S)_c$ becomes less accurate at low sulphur contents. In the author's formulation, $(Mn/S)_c = 2.7$ is for the free-machining steel with 0.3% S and 200 ppm O; this compares well with the ratio 3.5 estimated from practical observations of the ductile/brittle transition in the hot rolling of free-machining steels.[15] An average value of $(Mn/S)_c = 3.0$ is used in marking the region of hot shortness in Fig. 10.9.

It should be emphasised that the foregoing evaluation of $(Mn/S)_c$ is for steel (i) in the as-cast condition and (ii) in which the precipitation of MnS occurs under the conditions of local equilibrium in the solidifying interdendritic liquid. In the case of rapid rates of solidification, as in continuous casting, the local reaction equilibrium may not prevail and some liquid 'FeS' may precipitate in the last liquid to freeze, resulting in low hot ductility even though the ratio Mn/S in the steel $> (Mn/S)_c$. However, upon annealing at a high enough temperature, the nonequilibrium 'FeS' will be converted to MnS, resulting in the recovery of hot ductility, as demonstrated by several examples in the work reported by Lankford.[16]

## 10.5 CaS FORMATION DURING SOLIDIFICATION

The sulphide shape control by the calcium treatment of steel discussed in section 9.10.4 is another consequence of microsegregation and reactions occurring in the interdendritic liquid.

During the calcium treatment of Al-killed low-sulphur steels, the primary reaction is the conversion of alumina inclusions to molten calcium

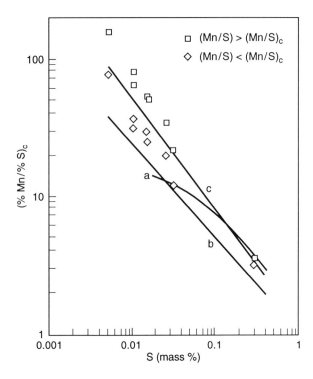

Fig. 10.10   The critical ratio $(\%Mn/\%S)_c$ as a function of sulphur content of steel. *Calculated lines: a* Ref. 8, *b* Ref. 12, *c* Ref. 13; *Experimental points:* □, ◇ Ref. 14.

aluminates containing a small amount of sulphur. As is seen from the reaction equilibrium data in Fig. 6.11, the equilibrium sulphur distribution ratio (S)/[S] for the reaction with calcium aluminate, increases with a decreasing temperature. As the steel temperature is decreasing from the end of calcium treatment to the start of casting, the aluminate inclusions in the melt will pick up sulphur as demonstrated in Fig. 10.11 for steels containing 0.01 and 0.06% Al, 10 and 40 pm S, 20 and 40 ppm O (total). The reaction with the aluminate inclusions shown below

$$(CaO) + 2[S] + [Mn] + \tfrac{2}{3}[Al] \rightarrow (CaS \cdot MnS) + \tfrac{1}{3}(Al_2O_3)$$

will continue in the interdendritic liquid during the progress of solidification, as depicted in Fig. 10.11. The lower the sulphur content and higher the aluminium, oxygen (as $Al_2O_3$) and calcium contents (as aluminate inclusions), the more extensive will be the sulphide shape modification, i.e. suppression of MnS precipitation. As shown in Fig. 10.12, the sulphide morphology modification is enhanced by manganese in the steel, particularly at lower levels of Ca and higher levels of S.

Steel in the tundish usually contains less than 25 ppm of Ca and O as calcium aluminate inclusions. Therefore, to achieve an acceptable level of sulphide shape modification, i.e. little or no MnS precipitation, the sulphur

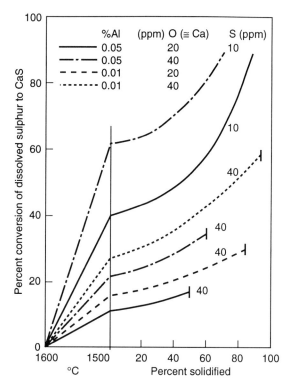

Fig. 10.11   Calculated conversion of dissolved sulphur to CaS on dispersed particles of calcium aluminate during cooling and solidification of steel.

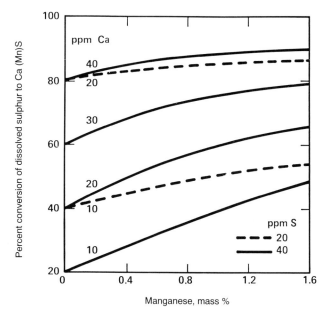

Fig. 10.12   Calculated conversion of sulphur in steel to Ca (Mn)S during solidification of the steel containing indicated amount of Ca, S and Mn.

content of the steel should be less than 30 ppm with Ca/S ratio of at least 1:1, or preferably somewhat higher.

## 10.6 Causes and Effects of Nitride Precipitation in Continuous Casting

The microalloyed HSLA steels are susceptible to the formation of transverse cracks on the surfaces of continuously cast slabs. This serious problem in the manufacture of HSLA plate steels (for line pipes, pressure vessels, and offshore oil constructions) has been investigated extensively since the late 1970s. For background information, reference may be made to a series of papers published in conference proceedings, *HSLA Steels Technology and Applications*[17] and *Vanadium in High-Strength Steel*[18].

In continuous slab casting of crack-sensitive steels, the transverse cracks are seen mostly on the loose surface (top surface) of the slab because the top surface of the slab is under tension during unbending at the straightener. The transverse cracks initiate at the valleys of deep oscillation marks and propagate below the surface along the austenite grain boundaries. As would be expected, the surface cracks are always filled with scale and show internal oxidation. The subsurface intergranular cracks, caused by the sulphide, nitride, and carbonitride precipitates at the prior austenite grain boundaries, are neither oxidised nor related to oscillation marks.

### 10.6.1 hot ductility

The initiation of transverse cracks at oscillation marks and their propagation below the surface of the slab is a manifestation of low-hot ductility of the steel at a temperature when the slab is being straightened. For a better understanding of the mechanism of formation of surface cracks in the continuous casting of slabs, numerous studies have been made of the effects of temperature and steel composition on the hot ductility of HSLA steels using the Gleeble hot-tensile test machine. In most tests, the steel sample is given a homogenising treatment at a temperature of 1300 to 1350°C, cooled to the test temperature in the range 600 to 1200°C and subjected to tensile test at the strain rate of about $3 \times 10^{-3}s^{-1}$, which approximates that experienced by the slab during bending and straightening in the strand.

As typical examples, some of the test results of Mintz and Arrowsmith[19] are shown in Fig. 10.13; similar results were obtained by many other investigators. The low ductility trough associated with the austenite $\rightarrow$ ferrite transformation at temperatures of 850 to 700°C, as noted in plain iron-carbon alloys,[20] extends to higher temperatures with the addition of

Al and/or Nb, even though the microstructure is fully austenitic. This effect of Al and Nb is, of course, due to the precipitation of AlN and Nb(C,N) at the austenite grain boundaries, as substantiated by numerous studies using electron microscopy and electron diffraction techniques. The papers cited in References 19 through 26 are typical examples of such studies. Even in a much earlier study by Lorig and Elsea,[27] in the mid-1940s, it was noted that the precipitation of AlN at the primary austenite grain boundaries was one of the principal causes of intergranular fracture in carbon and low-alloy steel castings.

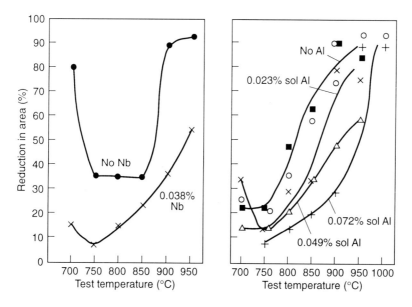

Fig. 10.13   Effects of Nb and Al on hot ductility of HSLA steels after solution treatment for 5 min. at 1330°C. From Ref. 19.

Both precipitates of AlN and Nb(C,N) at the austenite grain boundaries have similar deleterious effects on hot ductility. In fact, any fine precipitate at the austenite grain boundaries will lower the hot ductility of the steel. A rationale for the effect of precipitates on hot ductility may well be attributed to Hasebe,[28] who concluded from his earlier studies that the hot ductility derived from hot tensile tests is dependent on the effectiveness by which precipitates reduce the mobility of austenite grain boundaries.

It is now a well established fact that the low hot ductility trough in the C–Mn–Si steels, low in Al content, is associated with the austenite → ferrite transformation. The low hot ductility, resulting in intergranular fracture, is caused by strain concentration at the austenite grain boundaries accompanying the precipitation of the thin films of the softer ferrite phase at the grain boundaries.

In the case of the HSLA steels, the low hot ductility associated with intergranular fracture is caused by the grain-boundary precipitation of nitrides and carbonitrides which reduces grain-boundary mobility, leading to growth of voids around the precipitates, thus facilitating intergranular crack propagation, even when the microstructure is fully austenitic. As the concentrations of Al, Nb, C and N increase, the precipitation of nitrides and carbonitrides begin at higher temperatures, resulting in low hot ductility at temperatures well above the $A_3$ transformation temperature, hence widening the temperature range of low hot ductility.

There is always an impurity enrichment at the valleys of oscillation marks where the nitrides and carbonitrides will precipitate, causing local low hot ductility. When the oscillation marks are deep, the local rate of heat transfer will be reduced, resulting in blown grains below the depression, thus causing further local weakening of the surface layer. Noting that a deep oscillation mark increases the stress concentration due to the notch effect of the depression during straightening of the slab, the surface cracks initiate at the valleys of deep oscillation marks. Nevertheless, the primary cause of crack initiation and propagation is the precipitation of nitrides and carbonitrides concentrated at surface depressions and along the austenite grain boundaries. Without precipitation of nitrides and carbonitrides, surface ripples alone cannot cause transverse cracks.

The process variables, such as rate of cooling and/or rate of straining, affect the particle size and rate of precipitation of nitrides and carbonitrides which, in turn, govern the temperature range of the low hot ductility trough. For example, Mintz and Arrowsmith found that cooling the steel at a faster rate to the Gleeble test temperature, gave significantly less precipitation of Nb(C,N) in the matrix and a much finer precipitation of Nb(C,N) at the austenite grain boundaries, resulting in a lower hot ductility. They also showed that in the Gleeble tension tests, a slower strain rate widened the temperature range of low hot ductility.

In their study of the influence of prior precipitation on hot ductility, Wilcox and Honeycombe[24] found that the temperature range of low hot ductility in the HSLA steels increased when the steel sample was given an intermediate heat treatment of 15 minutes at 850°C, near the $A_3$ temperature; their experimental results are reproduced in Fig. 10.14. This effect of intermediate heat treatment at a lower temperature is due to an increase in the volume fraction of the precipitates of AlN and Nb(C,N) on reheating, resulting in the extension of the low hot ductility trough to higher temperatures. It follows from these observations that a temperature cycle near the slab surface in the secondary cooling zone below the mould will enhance the precipitation of nitrides and carbonitrides, consequently increasing the chances of formation of surface cracks at the slab straightener.

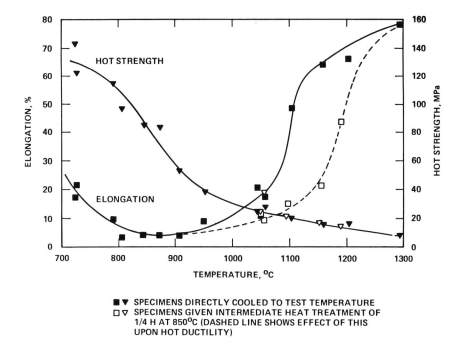

Fig. 10.14   Hot ductility (■, □) and hot strength (▼, v) of HSLA steel after solution treatment for 30 min. at 1300°C, the steel composition: 0.24 C, 1.20 Mn, 0.07 Nb, 0.06 Al, 0.01 N. From Ref. 24.

It should be noted that the intergranular precipitates are not readily seen under electron microscopy on replicas taken from transverse sections across the intergranular cracks because of the two-dimensional nature of the intergranular precipitate. With the technique developed by Wilcox and Honeycombe, direct carbon extraction replicas taken from the intergranular fractures were used to study grain-boundary precipitation.

In accord with the temperature range of low hot ductility caused by precipitation and $\gamma \rightarrow \alpha$ transformation, the depth of transverse cracks will depend on the temperature gradient below the surface of the slab. That is, the cracks will propagate to a certain depth where the temperature is high enough for the steel to be in the ductile range and/or where the tensile stress field becomes too low. It should also be noted that the transverse cracks may propagate deeper below the surface during cold scarfing to remove surface ripples because of the steep temperature gradient generated during scarfing, resulting in thermal stresses below the slab surface.

Tramp elements, such as Cu, Sn, Sb and iron-rich Fe(Mn)S accumulating at the austenite grain boundaries because of microsegregation during solidification, are also responsible for low hot ductility of austenite, particularly in the as-cast condition.

Then there is the effect of grain size on the hot ductility of austenite. As demonstrated by Ohmori and Kunitake[29], for a given mode of matrix strengthening mechanism, the elongation in tensile testing is determined by the prior austenite grain size and is linearly proportional to the inverse of the prior austenite grain size. In a recent study, Maehara *et al.*[30] found that the austenite grain size in the as-cast structure depends largely on the carbon content of the steel, the maximum grain size being in the 0.10 to 0.15% C region where the hot ductility is low. This effect of carbon on the hot ductility of austenite in the as-cast condition is in addition to the well-known carbon dependency of surface-cracking susceptibility in continuously cast slabs,[31–33] which is due to the δ-ferrite → austenite phase transformation in low-alloy steels containing 0.09 to 0.16% C.

Mintz and Arrowsmith[19] noted that slabs from casts (low Al content) which gave no plate rejections had a course distribution of Nb(C,N) at the austenite grain boundaries (Fig. 10.15a); interparticle spacing and mean particle size of ~140 and 40 nm, respectively. Slabs from casts (high Al content) which had high plate rejections were found to have a fine dense boundary precipitation of Nb(C,N) (Fig. 10.15b); interparticle spacing and mean particle size of ~60 and 14 nm, respectively. Similar observations were made by Funnell[23] with the precipitation of AlN in low-alloy steels; smaller AlN particles at the austenite grain boundaries, causing poorer hot ductility than coarser particles. The observed particle-size effect on hot ductility is in general accord with the expected particle-size effect on the grain-boundary mobility: the coarser the particle size and greater the interparticle spacing the lesser is the reduction in the grain-boundary mobility, i.e., the higher the hot ductility.

The practical significance of these findings is that the HSLA steels may be rendered less susceptible to cracking during hot deformation by controlling the precipitation of nitrides and carbonitrides so that coarse precipitates are formed at the austenite grain boundaries.

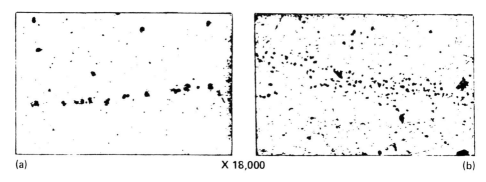

(a)                              X 18,000                              (b)

Fig. 10.15   Precipitation of Nb(C, N) at the austenite grain boundaries in HSLA steels: (a) coarse precipitate in the cast (0.021% Al) having no plate rejections; (b) fine precipitate in the cast (0.036% Al) with high rejection level. From Ref. 19.

10.6.2 MODES OF NITRIDE AND CARBONITRIDE PRECIPITATION

The precipitation of nitrides and carbonitrides plays a key role in the grain refinement of HSLA steels through controlled hot-rolling and ther-momechanical treatment. As part of the development work on HSLA steels, numerous studies have been made of the solubilities of nitrides and carbides in low-alloy steels. A detailed information is given in a previous publication by the author[34] on the reassessed values of the equilibrium solubility products of nitrides and carbides; these are summarised in Table 4.6.

In section 4.5, two examples were given of the equilibrium precipitation of AlN and Nb(C,N) in Fig. 4.31: (a) precipitation in homogeneous aus-tenite and (b) precipitation in the solute enriched regions which would be the case for the as-cast structure. However, the nitrides and carbonitrides do not precipitate readily during cooling of the steel casting or during cooling of the steel after the solution treatment. Gladman and Pickering[35] for example, have shown that the steel containing 0.08% Al and 0.012% N, after the solution treatment at 1350°C, had to be cooled at a rate of 1 to 2 °Cmin$^{-1}$ before any appreciable precipitation of AlN occurred below 900°C at which the supersaturation ratio, as defined below, is about 53.

$$\text{Supersaturation ratio} = \frac{[\%X]\,[\%Y]}{\{[\%X]\,[\%Y]\}_e}$$

where the subscript $e$ indicates the equilibrium solubility product at a given temperature.

In hot ductility tests, steel samples are cooled to the test temperature at relatively high rates, e.g., >50°C/min, yet there is a grain-boundary pre-cipitation of nitrides and carbonitrides which results in low hot ductility. It has long been concluded from these observations that precipitation from the supersaturated solution is strain induced and occurs during the me-chanical test to measure ductility at a particular temperature.

The strain-induced precipitation from the supersaturated solution is shown schematically in Fig. 10.16 with respect to the equilibrium pre-cipitation. The equilibrium precipitation occurs only in the isothermal treatment of steel for which the treatment time is longer at lower tempera-tures, e.g. about 30 minutes at 1100°C and 72 hours at 850°C. As men-tioned earlier, Mintz and Arrowsmith found that a strain-induced precipitation occurs at higher temperatures, i.e. lower supersaturation ratios as the strain rate decreases. In continuous casting, the strain exerted on the slab surface during bending in the strand would be adequate for the onset of a precipitation of nitrides and carbonitrides, leading to crack formation at the straightener. The temperature cycle at and near the slab surface caused by nonuniform water-spray cooling in the strand, will also increase the extent of strain-induced precipitation, enhancing the crack susceptibility of the steel.

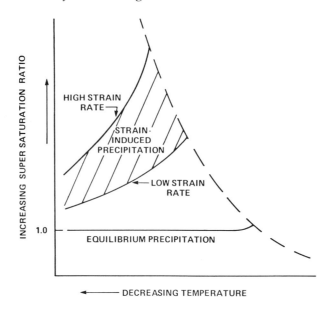

Fig. 10.16 Nitride and carbonitride precipitation from supersaturated solution in austenite.

The supersaturation of nitrides and carbonitrides in austenite during cooling of the casting can be prevented, by having in the solidified steel a certain volume fraction of appropriate inclusions, upon which the nitrides and carbonitrides grow with little or no supersaturation. A small amount of Ti addition to the steel is expected to bring about coarsening of the grain-boundary precipitation of AlN and Nb(C,N).

10.6.3 INTERDENDRITIC PRECIPITATION OF TiN

The solubility product of TiN in low-alloy liquid steel is $[\%Ti][\%N] = 6.15 \times 10^{-4}$ at 1500°C, which is taken to be the average liquidus temperature of low-alloy steels. The calculated TiN precipitation in the interdendritic liquid is shown in Fig. 10.17 for steel containing 0.02% Ti and 60 ppm N.

About 55 percent of the nitrogen is converted to TiN precipitate at 99 percent local solidification. Upon complete solidification, most of the remaining N in solution will precipitate as TiN, provided $\%Ti/\%N \geq 3.42$. The volume percent of TiN precipitated in the interdendritic liquid, which subsequently becomes the austenite grain boundary, is shown in the lower diagram in Fig. 10.17.

The particles of TiN seated at the austenite grain boundaries will be the sites upon which $NbC_{0.87}$ will precipitate during cooling of the steel, in accord with the equilibrium precipitation as shown in Fig. 4.31b for

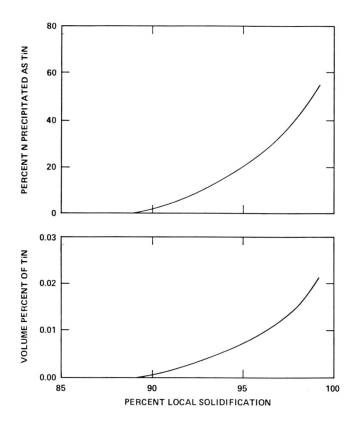

Fig. 10.17   Calculated TiN precipitation in the interdendritic liquid during solidification of steel containing 0.02% Ti and 60 ppm N.

Nb(C,N). Although $NbC_{0.87}$ is soluble in TiN, because of the low inter-diffusivities in the carbonitrides and relatively fast rate of cooling of the steel, the $NbC_{0.87}$ will precipitate epitaxially on the TiN particles. By this means of enforcing the equilibrium precipitation of $NbC_{0.87}$ at higher temperatures, resulting in coarser precipitates at the austenite grain boundaries, the low hot ductility trough will be shifted to the lower temperature range where $\gamma \rightarrow \alpha$ transformation occurs, i.e. about 850 to 700°C. Now if the surface temperature of the slab at the straightener is maintained at a temperature of either above 900°C or below 700°C, the transverse surface or subsurface cracks should not develop during straightening of the slab.

Similar to TiN, the solubility for ZrN in liquid steel is low: [%Zr][%N] = $1.9 \times 10^{-3}$ at 1500°C. Therefore, the precipitation coarsening can be achieved also with the addition of Zr. However, the required minimum mass ratio %Zr/%N = 6.52 is about twice that with Ti. Furthermore, the nozzle blockage becomes a serious problem in continuous casting of steel containing Zr. For these reasons Ti would be the preferred addition to achieve precipitate coarsening.

If the Al and N contents of steel are sufficiently high, there will be AlN precipitation in the interdendritic liquid, providing sites for the growth of other nitrides and carbonitrides during cooling of the steel, as noted in previous studies.[19,36] Because of other adverse effects however, high levels of Al and N in the steel cannot be tolerated.

With the titanium fixation of nitrogen as TiN, most of the aluminium will remain in solution in the steel, and the grain refinement during controlled rolling will be primarily via the solution and reprecipitation of niobium-vanadium carbides containing some TiN. The primary role of Al in such steels is in the deoxidation of liquid steel, for which 0.02 to 0.03% Al (dissolved) would be adequate.

### 10.6.4 CRITICAL MICROALLOYING ELEMENTS IN HSLA STEEL

There are many publications on the recent developments of HSLA plate steels for continuous casting, reporting similar results and recommendations for improved practices. In addition to the papers cited so far, reference may be made also to the papers by McPherson *et al.*[37,38] and by Coleman and Wilcox.[39]

Concensus is that the slab-rejection rate due to transverse cracks is reduced by decreasing the contents of Al, Nb and N in steel containing <0.09% C. The addition of Ti up to about 0.02% has been noted to reduce the crack sensitivity of steel. Microalloying of HSLA steels are usually in the ranges shown below.

> 0.06  to 0.08% C
> 0.02  to 0.03% Al
> 0.015 to 0.025% Nb (if the steel is normalised)
> 0.008 to 0.015% Nb (if the steel is quenched and tempered)
> 0.06  to 0.08% V
> 0.01  to 0.02% Ti
>         < 50 ppm N
>         <  5 ppm B

As for the casting practice, there should be uniform cooling of the slab surface in the spray zone, also the slab surface temperature at the strand straightener should be either above 900°C or below 700°C, to ensure that no surface and subsurface transverse cracks are formed in the continuous casting of high-strength, low-alloy steels.

## REFERENCES

1. B. CHALMERS, *Principles of Solidification*, John Wiley & Sons, Inc. New York, 1964.
2. W.A. TILLER, K.A. JACKSON, J.W. RUTTER and B. CHALMERS, *Acta Metallurgica*, 1953, **1**, 428.
3. W.C. WINEGARD and B. CHALMERS, *ASM*, 1954, **46**, 1214.
4. M.C. FLEMINGS, *Solidification Processing*, McGraw-Hill, New York, 1974.
5. H.D. BRODY and M.C. FLEMINGS, *Trans. Met. Soc. AIME*, 1966, **236**, 615.
6. E.T. TURKDOGAN, *Trans. Met. Soc. AIME*, 1965, **233**, 2100.
7. E.T. TURKDOGAN and R.A. GRANGE, *J. Iron and Steel Inst.*, 1970, **208**, 482.
8. E.T. TURKDOGAN, *Fifth Inter. Iron and Steel Congr.*, (Process Tech. Proc.) 1986, **6**, 767.
9. E.T. TURKDOGAN, *J. Metals*, 1967, **19**(1), 38.
10. B. HARKNESS, A. NICHOLSON and J.D. MURRAY, *J. Iron and Steel Inst.*, 1971, **209**, 692.
11. D. BURNS and J. BEECH, *Ironmaking and Steelmaking*, 1974, **1**, 239.
12. H. NAKATA and H. YASUMAKA, *Trans. Iron and Steel Inst. Japan*, 1986, **26**, B–98; *Tetsu to Hagané*, 1990, **76**, 376.
13. G.A. DE TOLEDO, O. CAMPO and E. LAINEZ, *Steel Research*, 1993, **6**, 292.
14. E. SCHMIDTMANN and F. RAKOSKI, *Arc. Eisenhüttenwes*, 1983, **54**, 357.
15. S.Y. OGAWA, T.B. KING and N.J. GRANT, *Trans. AIME*, 1962, **224**, 12.
16. W.T. LANKFORD, *Metall. Trans.*, 1972, **3**, 1331.
17. *HSLA Steels Technology and Applications*, (M. Korchynsky, ed.), ASM, Metals Park, Ohio, 1983.
18. *Vanadium in High Strength Steel*, VANITEC, London, 1979.
19. B. MINTZ and J.M. ARROWSMITH, *Met. Tech.*, 1979, **6**, 24.
20. T. NOZAKI, J. MATSUNO, K. MURATA, H. OOI and M. KODAMA, *Trans. Iron Steel Inst. Japan*, 1978, **18**, 330.
21. M. HATER, R. KLAGES, B. REDENZ and K. TAFFNER, *Steelmaking Conference*, 1973, **56**, 202.
22. K. KONOSHITA, G. KASAI and T. EMI, in *Solidification and Casting of Metals*, 268–274, 1979, London, The Metals Society.
23. G.D. FUNNELL, in *Hot Working and Forming Processes*, 104–107, 1979, London, The Metals Society.
24. J.R. WILCOX and R.W.K. HONEYCOMBE, in *Hot Working and Forming*, 108–112, 1979, London, The Metals Society.
25. G.D. FUNNELL and R.J. DAVIES, *Met. Tech.*, 1978, **5**, 150.
26. L. ERICSON, *Scand. J. Metall.*, 1977, **6**, 116.
27. C.H. LORIG and A.R. ELSEA, *Trans. AFS*, 1947, **55**, 160.
28. S. HASEBE, *Tetsu-to-Hagané*, 1962, **48**, 1575.
29. Y. OHMORI and T. KUNITAKE, *Metal. Sci.*, 1983, **17**, 325.
30. Y. MAEHARA, K. YASUMOTO, Y. SUGITANI and K. GUNJI, *Trans. Iron Steel Inst. Japan*, 1985, **25**, 1045.
31. S.N. SINGH and K.E. BLAZEK, *J. Metals*, 1974, **26**, (10), 17.
32. A. GRILL and J.K. BRIMACOMBE, *Ironmaking Steelmaking*, 1976, **3**, 76.
33. T. SAEKI, S. OOGUCHI, S. MIZOGUCHI, T. YAMAMOTO, H. MISUMI and A. TSUNEOKA, *Tetsu-to Hagané*, 1982, **68**, 1773.

34. E.T. TURKDOGAN, *Iron and Steelmaker*, 1986, **16**(5), 61.
35. T. GLADMAN and B.F. PICKERING, *J. Iron Steel Inst.*, 1967, **205**, 653.
36. B.C. WOODFINE and A.G. QUARRELL, *J. Iron Steel Inst.*, 1960, **195**, 409.
37. N.A. McPHERSON and R.E. MERCER, *Ironmaking and Steelmaking*, 1980, **7**, 167.
38. N.A. McPHERSON, A.W. HARDIE and G. PATRICK, *Inter. Iron Steel Trans.*, 1983, **3**, 21.
39. T.H. COLEMAN and J.R. WILCOX, *Mater. Sci. Tech.*, 1985, **1**, 80.

# Index